Personal Media and Everyday Life

DOI: 10.1057/9781137446466.0001

Other Palgrave Pivot titles

Nikolay Anguelov: Policy and Political Theory in Trade Practices: Multinational Corporations and Global Governments

Sirpa Salenius: Rose Elizabeth Cleveland: First Lady and Literary Scholar

Sten Vikner and Eva Engels: Scandinavian Object Shift and Optimality Theory

Chris Rumford: Cosmopolitan Borders

Majid Yar: The Cultural Imaginary of the Internet: Virtual Utopias and Dystopias

Vanita Sundaram: Preventing Youth Violence: Rethinking the Role of Gender and Schools

Giampaolo Viglia: Pricing, Online Marketing Behavior, and Analytics

Nicos Christodoulakis: Germany's War Debt to Greece: A Burden Unsettled

Volker H. Schmidt: Global Modernity. A Conceptual Sketch

Mayesha Alam: Women and Transitional Justice: Progress and Persistent Challenges in Retributive and Restorative Processes

Rosemary Gaby: Open-Air Shakespeare: Under Australian Skies

Todd J. Coulter: Transcultural Aesthetics in the Plays of Gao Xingjian

Joanne Garde-Hansen and Hannah Grist: Remembering Dennis Potter through Fans, Extras and Archives

Ellis Cashmore and Jamie Cleland: Football's Dark Side: Corruption, Homophobia, Violence and Racism in the Beautiful Game

Ornette D. Clennon: Alternative Education and Community Engagement: Making Education a Priority

Scott L. Crabill and Dan Butin (editors): Community Engagement 2.0? Dialogues on the Future of the Civic in the Disrupted University

Martin Tunley: Mandating the Measurement of Fraud: Legislating against Loss

Colin McInnes, Adam Kamradt-Scott, Kelley Lee, Anne Roemer-Mahler, Owain David Williams and Simon Rushton: The Transformation of Global Health Governance

Tom Watson (editor): Asian Perspectives on the Development of Public Relations: Other Voices

palgrave▸pivot

Personal Media and Everyday Life: A Networked Lifeworld

Terje Rasmussen
University of Oslo, Norway

DOI: 10.1057/9781137446466.0001

First published 2014 by
PALGRAVE MACMILLAN

Palgrave Macmillan in the UK is an imprint of Macmillan Publishers Limited, registered in England, company number 785998, of Houndmills, Basingstoke, Hampshire RG21 6XS.

Palgrave Macmillan in the US is a division of St Martin's Press LLC, 175 Fifth Avenue, New York, NY 10010.

Palgrave Macmillan is the global academic imprint of the above companies and has companies and representatives throughout the world.

Palgrave® and Macmillan® are registered trademarks in the United States, the United Kingdom, Europe and other countries

ISBN: 978-1-137-44647-3 EPUB
ISBN: 978-1-137-44646-6 PDF
ISBN: 978-1-137-44645-9 Hardback

This book is printed on paper suitable for recycling and made from fully managed and sustained forest sources. Logging, pulping and manufacturing processes are expected to conform to the environmental regulations of the country of origin.

A catalogue record for this book is available from the British Library.

A catalog record for this book is available from the Library of Congress.

www.palgrave.com/pivot

DOI: 10.1057/9781137446466

Contents

DOI: 10.1057/9781137446466.0001

DOI: 10.1057/9781137446466.0001

1
Introduction: Personal Media

Abstract: *This chapter introduces the central argument of this book and its key terms like 'personal media', 'personalisation' and the 'lifeworld'. It exemplifies the unwritten history of personal media by focusing on media of writing, talking and watching, and also between their primary functional modes that are called* orientation, interaction, presentation and archiving.

Keywords: communication theory; digital media; everyday life; lifeworld; media theory; personal media; social capital; sociology

Rasmussen, Terje. *Personal Media and Everyday Life: A Networked Lifeworld*. Basingstoke: Palgrave Macmillan, 2014. DOI: 10.1057/9781137446466.0002.

This little book addresses the convergence of the mediated and the (inter) personal in personal media of everyday life. From the convergence of new multifunctional personal media and the lifeworld, a series of new questions arises that this book only begins to address, concerning sociability and social integration as well as power and control.

The book introduces three main underlying arguments: First, that what I call personal media has its own history and its own sociology, distinct from the mass media. Second, that the concept of the lifeworld helps to understand meaning and power in contemporary everyday life in western societies, and, yet, that it needs revision according to the new reality of personal and mobile media. And third, that the overarching values of society give way to more or less mediated social networks as generators for social cohesion and integration.

Unlike the mass media, personal media favour interpersonal contact with family members, friends, colleagues and others we know. Our 'face time', as Rich Ling notes, is being extended and embroidered by mediated interaction. The notion of an everyday life refers to something we all experience, but is far from established as an analytical concept in social theory. Concepts like 'everyday life', 'the private sphere', the 'lifeworld' and 'domestic' tend to overlap, while they are also tied to different theoretical traditions. From a Marxist perspective, the private sphere is the space of reproduction and consumption. From a Weberian perspective, everyday life in the modern world remains the sphere of non-instrumental action. As everyday values and practices constitute a foundation of the understanding of technology presented here, we should develop a conception of the hermeneutics of everyday reality. We can derive such a conception by beginning with Husserl's lifeworld concept as it is developed by Gadamer, Schutz and Habermas, and continuing to sociological interpretations by Heller, Lefebvre, Goffman and Gullestad. A concept of 'everyday life' should keep the basic hermeneutic epistemology intact, while leaving out a too rigid and descriptive understanding of the term. Methodically, it should be able to guide empirical investigations and to make sense of specific day-to-day practices in a larger context of social integration.

Everyday life is a mixture of the unnoticed and inconspicuous on the one hand, and the partly strange and abstract arsenal of goods and services for consumption on the other. In his essay on the urban way of life, Simmel famously discussed this confrontation between the ordinary and modern life. The familiar world of daily life is continuously dealing with standardised and mass-produced objects and structurally planned

DOI: 10.1057/9781137446466.0002

environments. The media explosion of course is one particularly noticeable aspect of the last two decades. How does this disruption take place when the computer and the mobile are placed in the terrain vague between the trivial and the unfamiliar? Is it really a question of disruption or confrontation, or a smooth assimilation? In subtle ways, everyday processes adjust and absorb new technologies and media, as they must with regard to news and information. This has interested ethnographers and sociologists for a long time, as they stress the appropriation of domestic consumption (Gershuny), objects (Miller), TV (Morley, Lull, Silverstone), mobile (Ling, Prøitz) and other media. Sociologists and theorists like Henri Lefebvre, de Certeau, Bourdieu and others have attended to studies of the ordinary with various motives, most importantly because that is where life is lived, this is where the world presents itself, conditioned by class, gender, ethnicity and the brutal randomness of life.

The mobile and the laptop are being massively incorporated into people's lives, which cannot remain unchanged. The integration of digital media in expressive and instrumental operations in everyday life matters and practices is considerable. It is difficult to say how radical the implications of personal media are on our lives precisely because communication and its devices are fundamentally ingrained in nearly everything we do. But common sense signals to all of us that the changes are dramatic – not because they disrupt the humdrum of daily life, but because they change the way we do ordinary things. Just like people in Europe did one hundred years ago, we read, play, talk, shop, cook, write, enjoy, listen and travel every day, but the ways in which we do it change because new technologies invite us into more convenient avenues of practice. We are surrounded by voices, music and other sounds, texts, images and textures, but from very different sources than those of our parents' generation. For the last decades or so, everyday communication in particular is affected fundamentally deeper than when the telephone, the radio and TV were introduced. Web, webmail, Youtube, Facebook, Google and a variety of app-based services are now basic ingredients in our lives that reorganise our experiences and practices in time and space; they enable new social networks in new forms and genres.

Much of the changes are about the transfusion of power from social institutions like the school, the workplace, the church and the university, to markets, media, the home and the individual. Synchronisation is made more flexible by social institutions, which make the individual and the family much more the centre of the world. Individual movements and

DOI: 10.1057/9781137446466.0002

interactions in everyday life are now more in the hands of the individual. This rearrangement of power, giving private life a new dimension of flexibility that was unknown to ordinary families in the post-war era, has been a central nerve of modernisation since the 1970s. It implies more freedom of choice for individuals, within structural delimitations. It changes the lifeworld. The rigid regularity of Weber's 'iron cage' is turned into something softer and more flexible, or with Bauman, more 'liquid', however also more integrated into techno-social structures than ever in the history of modern society.

This is the contextual background for this book in conjunction with technological change. In the following, I address changes of everyday life in theoretical terms, and argue that we need a different conception of the lifeworld to contextualise social interaction with personal media. The lifeworld concept helps to make sense of daily life if it can account for actual changes in everyday life as related to individualisation of identity formation and personalisation of media. An understanding of the particular values, practices, habits and rituals of domestic life is vital to the understanding and use of new media practices. In this regard, I think we need a quite different notion of lifeworld than Habermas' version.

Once we acknowledge that personal media technologically and analytically belong to another world than the mass media, another media history and media sociology appears. This history does not start with Gutenberg's mass production of holy texts, followed by newspapers, film and broadcasting. Rather, the history of personal media probably begins with private notebooks in Greek and Roman antiquity, followed by letter correspondence through couriers and postal systems. It continues with electro-magnetic telegrams followed by telecommunications and a variety of innovations on the Internet and the Web. This history is not about audiences but about social relations in an extended space: on how individuals and groups interact with absent, and yet specific, others, and by necessity – with oneself.

The emergence of a national and international media industry from the 17th to the 21st century (printed books, the press, magazines, broadcasting etc.) was based on various forms of technological means that served as media, from the production (supply) side to the consumption (demand) side of the audiences. These technical media served as dissemination media, and constituted in their very structure a rupture between production and consumption, which prevented responses and input from the members of the audience. This paradox, the bridging

DOI: 10.1057/9781137446466.0002

and fencing, the dissemination and separation, which constructed and addressed audiences precisely, enabled, from Gutenberg on, an enormous international media industry. Because this industry created audiences who were impossible to actually observe, the industry had to rely on implied audiences.

However, since 1995 or so, and based on more than 150 years of telecommunications and 40 years of Internet communication, media forms have emerged in daily life that deviate radically from the paradoxes of the mass media. A remarkable wave of innovation in the area of personal communication, largely connected to the laptop, the web and the mobile, constantly offers their services to the modern individual. This is mostly evident with regard to the Internet. Due to the distributed end-to-end architecture of the net, new innovations constantly emerge in its periphery that increasingly rely on user-produced content in one way or another. In the 1990s, the Internet and hypertext converged in the web. Around 1998, a new generation of the web began (outside the news sites) with a wave of innovations, which much more actively integrate user-produced content. In the telecom sector, large-scale investments have allowed for a wave of innovations based on the upgraded mobile as terminal. Involved in what Weber called rationalisation has personalisation as one of its most evident trends characterising contemporary complex society.

In social research, the term 'personalisation' has been used primarily by Japanese researchers to describe the trajectory of telecom-based media from the business market to the household and the individual, and the transformation of the media during this process (see Kohiyama 2006, Okada 2006, Matsuda 2005). In Britain, 'domestication' caught on in the 1980s as a concept describing the introduction of various media into the household. Also in the 1980s, diffusion theorist Everett Rogers applied the term 'demassification' to account for the range of new computer-based media like electronic bulletin boards and email. By personalisation I mean the reorganisation of technology according to the expectations of persons. Kohiyama (2006, 71) argues that personalisation involves enabling access to the broadest possible range of Who, What, When, Where and How, as well as enabling people to specify and restrict access based upon their individual needs. Personalisation refers to devices and their design, such as smartphones, handsets and laptops, to personalised connectivity through passwords, credit cards, and other security and authenticity functions, and finally to personalised services,

DOI: 10.1057/9781137446466.0002

like shopping and downloading. While laptop-based communication was ahead of mobile-based communication, this gap is closed. Phones, tablets and laptops are rearranging their places in everyday practices continuously.

A central distinction thus needs to be made between personalisation of information and personalisation of media. While the first is a transformation that concerns all media, the latter is a distinct trajectory in media history. Personalisation of information relates to selection and specialisation of services and features in the mass media along with linking to other media to establish feedback channels. Most of this kind of personalisation relates to strategies in the press and broadcasting to appear 'closer' to the audience through a number of techniques, ranging from informal talk, studio audiences, call-ins, SMS-TV and so on. This is largely an effect of the endemic structural paradox of the broadcasting industry and the press, the separation from the audience that the industries want to reduce. From the 1980s, the 'audience' has been a concept in crisis, as much more complex models of individualised and segmented audiences have emerged that use broadcasting as well as digital media to create new hybrids of communication and consumption. For instance, in the beginning of the 1990s, the TV-companies started to implement measures to gain individual TV ratings, rather than household ratings.

This book addresses implications of the personalisation of media, which refers to the neglected history and sociology of mediated interaction in private life, from the private letter to the telephone, the personal homepage, email, instant messaging, photo-blogs, the mobile and social networking sites. The sender acts as an autonomous being, as in taking holiday snapshots, making phone calls or running a personal blog. And for the last decade or so, the personal aspect has been even more intensified since personal media now make us each personally addressable and more or less perpetually accessible (Ling 2008, 3).

Of course, the processes of personalisation of media and the explosive adoption of personal media cannot be explained by technological change alone, but also by changes in society, particularly increasing personal wealth, urbanisation, changing consumption and lifestyle patterns, including for instance that young people spend more time away from home.

Typically, the history of personalised media is the story of the transition of communication media from military, organisational, domestic and then to personal contexts. In contrast to the mass media, the distinction between sender and receiver is often not of prime analytical

DOI: 10.1057/9781137446466.0002

relevance, as in letter correspondence and telephone conversations. Even the distinction between the initiator and the responder may not be necessary in order to understand social relationships. Personalising of media relates to transportation infrastructure and telecommunications, rather than to the history of mass media from Gutenberg to Fox. Rather than entertainment, journalism and adverts, this kind of information is about personal notes, practicalities, downloading useful information, chat for no particular purpose, making appointments, and airing of personal opinions in blogs and Twitter and so on. Thus, what is here called personal media does not produce mass communication and is non-mass media. They rather make up lines or threads of social networks.

As noted, the extraordinary about our new personal media is that they have entered the entire realm of the ordinary in less than two decades. They have become part of our conventions and habits, which in return have made the media ordinary and indispensable. The ordinary refers to what we do in everyday life and particularly the ways we conduct everyday practices like shopping, reading the paper, chatting with the bus driver and friends and family, listening to music, watching TV, wearing clothes, spending holidays and cooking dinner. Although many of these practices are signals of lifestyles and self-expression, they are also the low-key life itself, what we do while making plans and producing memories. The ordinary is a topic for reality shows, which is frequently discussed (see Bell and Hollows 2005). Here, ordinariness particularly denotes the mundane practices of keeping in touch, fulfilling one's obligations, relaxing, searching for a recipe, enjoying something and socialising in the ongoing business of everyday life. The indispensable emerges for exactly the same reasons. In brief, what seems to take place in everyday life is a series of expansions or transfusions which will be addressed in the following:

- from reception to production
- from interpersonal context to interpersonal communication
- from messages to communication
- from audiences to networks
- from domestication to personalisation
- from audiences to Self and other
- from the history of mass media to the history of personal media.
- from the history of media to the history of mediated communication

DOI: 10.1057/9781137446466.0002

Writing, talking, watching

Personal media are not new and have always been connected to meaning-constituting sensorial practices, particularly writing, talking and watching. While we normally use our senses in subtle combinations, media tend to specialise or combine more rigidly. The manuscript is for writing or reading, and occasionally we admire its beauty. Monocles and telescopes are media for watching. They tend to condition one another: writing follows reading, and talking on the phone is combined with hearing. The history of personal media is partly a history of combining sensorial practices linked to time and space in new combinations. Often, one of the sensorial practices is prioritised, with other sensorial practices accompanying. For the sake of ordering personal media in a social history, I believe it makes sense to study the history of personal media according to the leading sensorial practices that they manipulate. There is a history of writing (reading), a history of talking (hearing) and a history of watching (talking, hearing). I return in more detail to this point in the following chapter.

Personal media cannot meet the general expectations that media research traditionally formulates about the mass media. Rather, they ought to be conceptualised as media for personal production and reception of information, usually generating communication. The major sociological significance regarding the smart mobile for instance, is not that it is mobile as such (mobility is an effect of wireless protocols, its main technical feature), but that the mobility makes the phone personal, more so than the personal computer ever was. It enables constant personal contributions to verbal or textual conversations. And the number of web-based features and apps for the mobile that have emerged since around 2008 stem from new software and transmission capacity that make the web into a personal medium for presentation, interaction and archiving.

User-produced content is the substantial part of the genres and communication of personal media. We speak of personal media (today, mostly based upon the laptop, the 'pad' and the smartphone as hardware platforms) as something distinct from mass media. Personal media are designed for private individuals and are used for communication with one or many others in public, private and semi-public settings, and in institutional as well as non-institutional contexts. They are not only individualised (as are many mass media) but also personalised, as

DOI: 10.1057/9781137446466.0002

evidenced by user-identities and password-protections. In this, they go further than what Raymond Williams called 'mobile privatisation', and what Roger Silverstone and associates called 'domestication'. Another defining feature of personal media is that they enable user-initiated co-production of media content. Information and utterances become integrated as in phone conversations, SMS-interactions or networking sites. User-production affords presentation of self-reflective meaning, and thus a state of being in the world. It enables ongoing contact with distant and distinct others, and helps to coordinate everyday life.

Modes of personal media

Proposals on what distinguish new media from the mass media have put forward a series of analytical distinctions (digital media vs. analogue media, passive vs. active, interactive vs. non-interactive, monological vs. conversational, symmetrical vs. asymmetrical and so on), which quite often remain on the level of specific technological variants. But the convergence much debated in the 1990s also includes convergence between the digital and the analogue, between the digital media and the mass media. This obfuscates a clear-cut distinction between groups of media on technical grounds. And even if we were able to distinguish new media technically from the traditional mass media, our business is rather to understand the interplay between, or integration of, media and social life. Of interest are the social configurations that the various media provide.

In personal media, I distinguish between four ordering modes (or functions), which I call orientation, presentation, interaction and archiving. Up until the 1990s, they were connected to distinct technological platforms, such as post- and phone-based media of interaction and web-based media of presentation (personal websites and web-diaries, blogs), and computer-based archives and databases. The orientation function made a giant step with GPS and could be combined with the three other platforms.

With digitalisation and network convergence, technical distinctions blurred. The term 'convergence' referred to the fact that several networks, corporations and devices serve multiple functions, particularly the mobile and the laptop, and increasingly the TV set. Writing letters may afford presentation, interaction and archiving. Services may be used for

DOI: 10.1057/9781137446466.0002

other tasks and purposes than they were designed for. The platforms are increasingly used interchangeably for all functions, even if we still live in a world where archetype media like the wristwatch, wall calendar, camera and the plain old telephone co-exist with digital converged media. The smartphone is used for web services, which is an inroad to mobile TV and online news. Also, a number of websites and blogs draw upon presentation, interaction and archiving functions. A number of networking sites, such as Facebook and LinkedIn as well as blogs, apply both presentation and interaction functions in the construction and reproduction of relevant social networks for various purposes.

Thus the four functions in personal media – orientation, presentation, interaction and archiving – could refer to four types of media were it not for the facts of convergence and divergence as effects of social change and media digitalisation. Although it becomes less and less relevant to distinguish between personal media of orientation, presentation, interaction and archiving, they remain as modes in networks and platforms, which indicate the leanings and biases of distinct media architectures. Moreover, they indicate the existence of historical and theoretical inroads, which help us to understand these new developments more in depth. The distinctions between orientation, interaction, presentation and archiving are important to maintain for both historical and analytical reasons.

Orientation functions refer to meaningful location and mobility in time and space, such as is evident with the camera, the diary, the clock, the planner and the wristwatch. The perception mode refers to information about where and when, one's own or other's location in time and space. Perception refers to the body of the agent. In this case, the person acquires information from media like the map, the compass, the microscope and the telescope, the watch, glasses, the hearing aid, the calendar and GPS.

The presentational mode refers to what, to directing and attracting other's attention to one's publication of ideas, thoughts, opinions and information, as for instance in letters, SMS and on personal homepages on the web, blogs and twitter. Presentation functions are to be found in media as letters, business cards, books, homepages and blogs, where the individual (or group) presents itself to the world.

Interaction functions are for interacting with oneself and other individuals and groups, such as with the personal letter, telephony, email and social media. The interaction mode refers to who, to media for social

DOI: 10.1057/9781137446466.0002

interaction between two or more persons, as with letter correspondence, the electro-magnetic telegraph, the phone, email, and social media. These media are often referred to as dialogical or two-way oriented, as opposed to one-way mass communication.

Archiving functions refer to the ability to store/save documents in a retrievable way, such as notebooks, letters, photo albums and databases. The archiving mode also refers to when, to material memory, to storage media used to preserve meaning for the future, to retrieve past information, as on tape, answering machines, laptop hard discs and other personal archives such as photo albums, Filofax and other personal filing systems. It also, as I will address in a later chapter, constitutes a method of forgetting, and thus a method of directing culture.

Autonomy and ambivalence

Features of communication technologies indicate that they enhance greater personal freedom and independence. They enable communication with friends, work-relations, information providers and so on, at the time and place decided by the individual agent. It seems possible to tailor information and communication to habits and routines of personal life. We can have our cake and eat it too: work, play, go fishing and skiing, drive a car or live in rural areas while still flexibly communicating and transmitting information with others. From this, one cannot but conclude that the individual is more in control of her own life than before. At least for the well-off western individual, the world consists of an abundance of social and technological systems of expertise and technology, with all kinds of needs and interests at our service.

This autonomy is of course an illusion. The only way to live a normal life is to make use of system-produced services in a stable and efficient fashion. The process may present itself as an option, but it also implies social control. Systems providing services and comfort necessary in modern life can rarely be ignored. In exchange, we renounce the knowledge and power to question, understand and evaluate decisions concerning our lives. Everyday life is to 'go on' in a practical and reflexive manner, by the use of system-produced support. Modern life is lived in the grey zones of what Weber called value-spheres (private life, politics, science, economy) with their corresponding roles. Individual pseudo-autonomy is reached through mobility in space and flexibility in time (Rasmussen 1996).

DOI: 10.1057/9781137446466.0002

Personal media play a decisive role in this, since they modify and stabilise everyday networks. Related to this is the argument that personal media are not technologies in a usual sense, as analysed by social philosophers from Karl Marx via Herbert Marcuse to Neil Postman. They are not simply standardised means and rules under the rationality of instrumentality and calculation. They may be media of power but whose power is not immediately clear. Their communicative dimension seems to make them evade the definition of technology as instrumental, which explains why personal media and everyday life easily adapt to one another. Their rationality is as much a reflexive and communicative kind, as a purposive or strategic. With the growing autonomy that follows the modern world of communication technologies, parallel forms of freedom and dependence emerge. To live a modern convenient life implies that identification as well as socialisation increasingly draws on media connected to large-scale systems. This implies freedom as well as vulnerability and risk, as when the credit card is not accepted, when the suitcase never appears on the baggage belt and when the computer ignores our commands. The paradoxical development of autonomy and dependence poses new challenges for the modern individual.

It is through the many good things that online personal media offer, most of all enhancing love, friendship and family relationships, that they also occupy our time, present mindless adverts and harvest and aggregate our data for their customers. As Edward Snowden explained to a world audience in the summer of 2013, the National Security Agency in the United States taps and processes our private communication. There are many things to worry about concerning some of the personal media. However, Sherry Turkle's worry is not among mine: 'On social networks, people are reduced to their profiles [...] We are increasingly connected to each other but oddly more alone: in intimacy, new solitudes' (Turkle 2011, 18–19). As long as online relations are mixed with offline relationships, this worry is exaggerated. The ones we text and talk and email with through our mobile or laptop tend to be people we often meet face to face. Yes, they are time-consuming, distracting and trivial, and yet also social and occasionally vital in the struggle for democracy. They may be used for harassment and in campaigns for social justice. It is a natural part of our rationality to seek out the most convenient ways for achieving our goals. Sometimes, the most convenient is what serves us in a long-term perspective, but sometimes it is not.

*   *   *

DOI: 10.1057/9781137446466.0002

In Chapter 2, I go into more detail concerning the history and evolution of personal media along the three sensorial categories presented in the introduction. One of the purposes is to demonstrate the diversity of personal media excluded from media history. Another purpose is to connect this history to the sociological perspective of individualisation. Important aspects of the media evolution took part in the definition of the person as a self-conscious and autonomous being. This is the 'mediatisation' of the human being as an individual and as a person that took place alongside other rationalisation processes examined in sociology, from Weber to Luhmann, Beck and Giddens, that created the self-reflective individual and (high)modern society. I address aspects of the media of writing, talking and of watching and illustrate their development historically. I argue with theorists like Giddens, Goffman, Bauman, de Certeau and particularly Foucault ('Technologies of the Self') that this development has formative power on the formation of identity and self-reflexivity of the individual. I address social relationships in everyday life as the application of a mode of 'extended familiarity' between intimacy and formality.

In Chapter 3, I introduce 'everyday life' as a sociological category and examine more specifically and critically the concept of the lifeworld. I follow the concept from phenomenological philosophy to sociology, and offer critical remarks on Jürgen Habermas' version of the concept, which points towards a revised notion of the lifeworld more compatible with a modern, mediated everyday life. I then link the concept of the lifeworld to more recent theoretical and empirical trends, particularly concepts of 'domestication' (Silverstone), another concept that needs to be updated according to recent social and technological change. I address domestication more specifically as 'personalisation'. I also connect to de Certeau's theory of strategy and tactics in everyday life in order to reach at a critical but bore network-like, 'thinner' (more individual-oriented) understanding of the lifeworld. With this chapter, the basics of the setting for sociological research on the uses of personal media have been set in an analytical sense.

Chapter 4 addresses media and communication research. It begins with assessing the role of interpersonal communication in mass communication research and concludes that mass communication models will not do in examining the use of personal media in current, everyday life. If communication at all, mass communication is a special type of communication that deviates from intuitive understandings of communication

DOI: 10.1057/9781137446466.0002

as some form of interaction or dialogue. Most types of personal media are true communication media. Therefore, we need to take a step back to the basic models on direct, unmediated communication. I briefly address the 'communicative turn' from Husserl to contemporary philosophy and sociological theory. Some communication-theoretical proposals are discussed that might position media studies and sociology of everyday life to make sense of communication in personal media as everyday communication. I particularly discuss N. Luhmann's communication theory in some detail in relation to actual media use. I argue that it serves examination of personal media well at the same time as it connects such use to more general sociological insights.

Chapter 5 takes the step from communication and genres to media technologies. It presents some reflections on the relevance of what has been called 'medium theory' in order to approach an understanding of personal media. I particularly revisit M. McLuhan's ideas, including his less known or understood 'Laws of media'. I then address some other medium theories inspired by McLuhan (Bolter & Grusin, Langer, Watzlavic, Meyrowitz and others) Medium theory, I argue, through its theoretical balance of technology and message, grasp the power of the technology and to what extent it can be 'tamed' or domesticated. I address both the discomforting aspects of dealing with virtual (simulated) realities and the convenience they offer in our sociality. I address the ongoing discussion both in research and daily life itself, about the relationship between the online and offline in everyday life. I illustrate with the very recent development of augmented media/situated simulations.

Chapter 6 examines the accumulation and dislocation of social capital in everyday life and the intermediary role of personal media, particularly social media. The concept of social capital I argue is sociology's main gift to the personal media researcher; it helps to get an a priori balanced view on mediated social relationships in everyday life, with an eye on power. I address various approaches on social capital (Bourdieu, Coleman, Putnam, Granovetter and Burt), and from this I define the main elements of social capital. Individuals (and groups) are embedded in social relations, but are not necessarily exercising pure purposive rationality. A balance between a conception of the 'undersocialised' and 'oversocialised' individual can be found, using the concepts of skills, social capital and social network. This is done in order to emphasise the structural and relational character of resources in everyday life, without ignoring individual abilities and the uniqueness of social situations.

DOI: 10.1057/9781137446466.0002

Thus the concept fills an analytical gap between individual practices and everyday social networks, whether face-to-face or mediated. I refer to research that illustrates the relevance of the concept in various situations and for various kinds of media, and towards the end I particularly address the problem of trust.

DOI: 10.1057/9781137446466.0002

2
Encircling the Person

Abstract: *This chapter goes into more detail concerning the history and evolution of personal media along the three sensorial categories presented in the introduction. It addresses aspects of the media of* writing, talking and watching *and argues with theorists like Giddens, Goffman, Bauman, de Certeau and Foucault that this development has formative power on the dynamics of identity and self-reflexivity of the individual. It presents social relationships in everyday life as the application of a mode of 'extended familiarity' between intimacy and formality.*

Keywords: communication theory; digital media; everyday life; lifeworld; media theory; personal media; social capital; sociology

Rasmussen, Terje. *Personal Media and Everyday Life: A Networked Lifeworld.* Basingstoke: Palgrave Macmillan, 2014. DOI: 10.1057/9781137446466.0003.

DOI: 10.1057/9781137446466.0003

It is certainly correct that mass media took part in the encircling of the modern mass society of electoral democracy, mass consumption and mass education. The story is normally reconceptualised in sociology as modernisation, rationalisation, differentiation and similar terms. In the shadows of this evolution of mass society, another trend also implicated the media. A glance at the history of personal media indicates that the nature of these media in mundane, everyday contexts up through the centuries and decades has told an underlying story connected to the theme of rationalisation as well, about how a long history of material means of communication has gradually 'staged' the human being as an autonomous individual. This chapter offers a few retrospective snapshots of the history of personal media to illustrate the evolutionary trend towards personalisation.

Unlike most histories of digital technologies, a personal media history stresses everyday unique communication rather than standardised production, and private and personal involvement rather than professional use. Here, I simply want to state the point that the long history of such media has been overshadowed by the relatively short history of mass media and mass communication on the one side, and the even shorter history of computers and telecommunications on the other side. While the first analytically excluded such media as the letter and the telephone, the latter focused on professional use. The hitherto unwritten history of personal media may lead to a richer understanding of past and present social life where such media play an increasingly prominent role. The history of personal interaction over distances adds to the understanding of long-term trends.

Media and their physical elements (quills, brushes, pencil, pen, paper, typewriter, keyboard, joystick etc.) are material things that culture use for the production of meaning. Whether manuscripts, printed documents, radio programmes or websites, they are dead things that can talk to us and for us through the invention of code and genre. They are therefore both dead and material, and human-related and social. We study the material aspect in order to see how different media, through their internal dynamics, have influenced different cultures and epochs, and we study the social aspects in order to see how genres and styles have endured across epochs and media shifts. They are neither neutral information vehicles, nor are they only the productive and seductive forces of history over which we have little influence.

DOI: 10.1057/9781137446466.0003

Certainly, a key aspect of many personal media is that they transcend local, place-bound contexts of practice. The mobile is used for private matters, often in public contexts, occasionally to the irritation for others. Typically, personal media tend to define new social contexts in and through their use, but increasingly the use of personal media is omnipresent and close to context-independent. Most personal media used for communication undermine established boundaries for social interaction. And new ones are created, such as coffee-places now often serving as office landscapes. To understand the media-daily life dynamic, it is therefore important not to define media contexts rigidly. The range of information and communication from the intimate to the very public appears as a continuum where even the involved parts are often not in accord with their status. The main point here is to highlight the diversity of mediated communication that is not produced by mass media organisations, and to suggest perspectives on how they influence everyday life.

To present a comprehensive history of personal media lies beyond the scope of this book. Here I can only indicate, with some elements as examples, the range and span of such a history. As indicated in the introductory chapter, we may observe the history of personal media according to the leading sensorial practices that they manipulate as sub-histories of writing (reading), talking (hearing) and watching (talking, hearing).

Writing/reading

While talking is the synthesis of social action and communication, writing is the synthesis of social action, communication and the media. It probably began with the papyrus, the grass-like plant that has grown in the Nile river delta for thousands of years. With papyrus, nature was transformed into materiality. The plant was sliced lengthwise into long strips and laid in rows dried in the sun. It was, however, paper that made writing into a social practice. Paper was invented in China in the second-century CE and was carried to Europe in the 12th century by Arab traders (Levy 2001, 9). In European medieval monasteries, paper production was based on the mashing of plant fibre, spread out in thin layers and dried. The sheets could then be written on with a quill or a brush. Even if paper was more fragile than parchment from skins, it was cheaper to produce. But because of the quality of skins, paper did not supplant skins completely until the end of the medieval ages. When

DOI: 10.1057/9781137446466.0003

animal skin was used, the hair was removed, then stretched and sanded. The surface was thin and extremely durable. Unlike papyrus, parchment could be marked on both sides (Levy 2001).

Of course, the printing press created great demand for paper. Paper was then made out of rags from old clothing made of cotton, linen or hemp. As late as the mid-19th century, with the invention of the steam engine, techniques for making paper out of wood became commercially viable, and laid the foundation for a society that relied more and more on documents of all forms. Already by the end of the 13th century, a written document could serve as a witness and statements of facts. With the invention of carbon paper (in the 1820s), paper came to work well with the general need for speed and type-written documents in the late 19th century. This again gave way to new filing methods, such as the Dewey Decimal System (Levy 2001, 68).

By that time, the private sphere needed written and typed communication such as receipts, prescriptions, travel guides, tickets, maps, memos, job announcements and passports. Collections of such everyday media are historical documents telling stories about prices, appearances and travels. Private life needed documentation. Collections of tickets inform about moments of travels and events. And greeting cards, business cards, and postcards are all examples of everyday genres that do not seem to disappear with digitalisation, simply because they provide convenient forms of interaction. Most of these forms of notes and cards, including the Christmas cards, came into use in the 18th century. These are modern phenomena that indicate a growing complexity of social settings, with the corresponding need for social rules.

In Japan, textual communication (writing) has been more preferred than mediated talk. The most popular internet service on the mobile is email, and text messages are used more than voice calls among young people (Okada 2006, 49), beginning with the display on the pager in the early 1990s, where the caller typed in the number he or she wanted the receiver to call. This initiated a learning curve applying small mobile means for textual communication. Other reasons for this are that voice was more expensive than text, and that, in most countries and more in Japan than elsewhere, norms reduce speaking on the phone on collective means of transport and in other crowded public places.

With News-groups and email, the Internet was a text-only medium up until the early 1990s, and still text is dominating, however in junction with other medium forms. With the web and the *blog*, new writing

DOI: 10.1057/9781137446466.0003

media appeared that could quickly reach a limited group of readers who are familiar with the author or share the author's interests. Features known from the personal homepage were combined with elements from the discussion forum and online news sites into a flexible personal medium for both presentation (individual control) and social interaction (Hodkinson 2007, 626). Whether the emergence of the blog and other personal media like the mobile is likely to encourage more individualistic developments of identity cannot be positively confirmed. And yet we may safely assume that individualism and personal media generally coincide in time and are interrelated and mutually reinforcing in complex ways. No doubt the individualistic zeitgeist in our times encourages innovation in the areas of personal media and of personalisation of mass media, which provide means for developing identity. There are clear signs that parts of the audiences of radio and TV, as well as users of collective digital media like user-groups and large-scale discussion forums, expand or migrate to a wide range of media that underline individuality and identity. For instance, Hodkinson (2007) found that, when interaction moved to blogs, communication became more distinctly individual both in the form of diversity of content, format of conversation and networks of 'friends'. Personal media 'expect' and 'encourage' personal style and content at the expense of group or mass cohesiveness. Blogs (through their link features and comment facilities) and the mobile (through their conversations and messages) encourage sociability in the shape of networks, a more flexible and less norm-influenced form of sociability.

We used to think about intimate relationships as *close*, but that was before chatlines, partylines and datelines emerged in the late 1980s. In such services, a dial up chat may lead to conversations with another individual, without revealing identities, which may lead to a face-to-face meeting. In many countries, this service began as a low-status activity, partly associated with prostitution. In contrast, the chat rooms on the Internet that were introduced in the late 1980s and reached high popularity when appearing as web-services in the early 1990s were often seen as valuable and serious alternatives for singles. Today the diversity of discussion forums is vast, covering most segments, interests and tastes. The seriousness was also higher for Instant Messaging Systems that appeared in the mid-1990s (ICQ, AOL, MSN, YIM) and provided chat rooms as well. Anonymity is a feature that in most face-to-face and mass media settings is unacceptable. With the Internet, anonymity and pseudonymity was turned into a force to boost new services, making anonymity and intimacy not adversaries but companions. Technological

DOI: 10.1057/9781137446466.0003

innovation along with the human taste for carnivals and masquerades allowed this. In modern 'single' society, such forums provided new sites for romance and intimacy.

Beginning around 2005, the mobile entered this laptop-dominated landscape of 'lonely anonymous hearts'. In many countries following Japan and South Korea, major Internet portals offered chat rooms and conversation possibilities for strangers on mobiles, such as Dodgeball, Playtxt and Livedating. Already in 1998, devices enabling teenagers to find each other when out were introduced in Japan. Lovegety, Coofy, ImaHima and Navigety were simple devices for locating friends and sending small messages (Tomita in Ito et al. 2006). Online and mobile communities dedicated to various interests appeared, similar to the ones already established on the laptop. The smaller screen and keyboard proved no serious drawback, at least not compared to the advantage of extended mobility. Also, when linked to GPS, the location of services and people were included in digital personal communication. This digital managing of physical place implied that offline and online worlds merged even more.

In 1996, the now ended service ICQ ('I seek you') was introduced that made the service freely available for anyone with Internet access. A number of Internet messaging (IM) services were introduced by AOL, Microsoft, Yahoo and others. IM grew out of the use of electronic bulletin boards and online services built on dedicated software from providers such as Prodigy and AOL, and also from the Internet activity in the 1990s in chat rooms implemented in web communities. IM had a number of characteristics that made it suitable for youth communication. IM was a medium for quick asynchronous interaction, usually between two persons, in contrast to Internet Relay Chat, which was normally group-oriented. IM enabled private 'rooms', and one could have several dialogues going simultaneously. But IM also contained possibilities for chat, file-sharing, talk, streaming content and so on. And contrary to email, it assumed constant presence, similar to text messaging, its successor. Contrary to texting, it provided information on whether the other is online, and so provided an additional element of social control. When entering IM, one immediately could see who among one's friends were logged on. The buddy-list was a document of nodes in one's personal networks. The IM could enable individual relationships and at the same time underline one's membership in a peer group. It is precisely this duality that made it suitable for children from around the age of ten.

DOI: 10.1057/9781137446466.0003

Particularly for the youngest, the importance of having a long buddy-list was an ongoing confirmation of popularity, as is the case with the phone list and the friends-list on Facebook.

With IM, one could have a sense of control with several persons in one's peer group. It could be used as a secondary (background) medium, while doing primary activities. In this medium, it was not the group but the me–you relationship in the context of a larger peer group that proved attractive. Also, the group had a sense of being together even if the members of the group were not actually communicating. A message saying one has to 'leave' for dinner with the parents confirmed togetherness. The privacy of the channels made them trusted and suitable for intimate talk. Still, research indicates that youths did not find IM particularly enjoyable (Kraut et al. 2002, 215). This however could simply be a sign of integration in everyday life. As with the phone, it seemed to be the mundane convenience offered by IM that made it into a very frequently used medium for a distinct cohort. Following ICQ, MSN Messenger and Yahoo! Instant Messenger appeared, but did not apply any open protocol. The proprietary nature of these services prevented interoperability and undermined their robustness. The results of the work of Internet Engineering Task Force Group on this (the IMPP-working group) came too late. After around ten years of existence, IM lost its attraction, due to the popularity (and universality) of text messaging and chat-functions at web-based social networking sites.

The first text message (SMS) was sent by NOKIA engineering student Riku Pihkonen in 1993 (Agar 2003, 177). Expectations are generally high that messages are responded quickly to keep the network sufficiently dense in order to keep the awareness of a chat alive. Not contributing would create speculations and disappointments. Studies of SMS use report on the pressure to interact and the worries that emerge if messages are not returned. SMS thus represents an immediate mode of communication, and works on the assumption that the others are already 'here' even if the connection is not kept open (Ito and Okabe in Kraut 241). What emerged was a social network of binary interaction that fostered a temporary social space of 'us'. The network created an impression of togetherness in a stripped or 'thin' way. Particularly for young people, this medium appeared as a 'place' for socialisation among peers. Text message sequences are media for direct or pure interaction in the sense that openings, greetings and conventions of politeness are considered redundant.

DOI: 10.1057/9781137446466.0003

Particularly notable about everyday text messaging ('texting') was that the meaningful unit of analysis was not the singular message but the ongoing sequence of exchanged messages between close friends during the day. The singular message is often only a line or queue in an ongoing chat. In communication terms, the SMS channel does not open and close for each message but is already open and accessible as in face-to-face talk. That is why delayed SMS-replies are considered rude or exceptional. They generate a private space that is filled with everyday half-baked utterances or expressions and signs, each with little significance, but important as constitutive elements in talk. No other medium comes closer to being assimilated in everyday talk in this way.

Writing with software

A sociological remark is apt here. To apply technologies of writing to talk in email, on blogs and on Facebook is not an altogether trivial matter. Today, power is to a large extent what the sociologist Talcott Parsons called *influence,* or even *ideological* power, and ideologies are mediated through writing. That is why so much energy is invested in the design and rhetoric of modern texts, in all media, in all genres. Writing is a careful practice receiving its social significance from its rhetoric as much as from its substance.

Ideological power lies with those who possess this technique of producing rhetorically efficient writing, in the shape of novels, political propositions, scientific publications, schoolbooks and web pages. It is a form of power, which changes the social hierarchy of society by lending social status to the scientist, the bureaucrat and the author of books, web pages and documents. The power is exercised on behalf of all of us when we act out our roles as family members and consumers who only read and talk.

The web page does not alter this constellation between readers and writers by changing the imbalance between written power and oral-reading powerlessness. It is, however, quite clear that the web already has triggered new ideas of participation and new updated forms of democratic activity. The questions are: how are these structures to be handled? Can the human character, the person's integrity be identified on Facebook walls? Can we disclose the ethos, the common sense of the competent individual behind the blog?

DOI: 10.1057/9781137446466.0003

The resources at play here are of two types: (1) the medium of writing that records experiences and tells the stories and (2) the hardware and software as instruments of writing that cut across whatever is reported through them. *Inscription,* in actor–network theory, belongs to the latter type. Inscription means the way artefacts embody patterns of use. The term describes how anticipations and restrictions are involved in the development and use of a technology.

Media technology puts its mark on social expressions by leaving certain distinct traces of such action. Methodologically, we need to go beyond what is written as a road to the writing itself, and further, to the changes in self-presentation as a consequence of technological structures. The mark of the technology is seen in the design. The use of communication technology is constituted by semantic and normative rules, power and skills, along with material resources applied by the actor in interaction processes.

Anthony Giddens (1991) defined *structure* as sets of rules and resources that both empower and constrain social action, and that tend to be reproduced by that action. Structure is the medium and outcome of action. Applying software like Dot.mac Homepage, Yahoo's homepage program, Powerpoint or Blogger implies that one adjusts one's practices to certain flexible, but also already existing, structures. We are involved with the 'acquisition of technology and its dispositions' as Pierre Bourdieu could have formulated it.

It's only about communication now

Among the first modern philosophers of the modern and autonomous Self, Rousseau pointed out that to become a subject both in terms of reason and in terms of emotions, one must stand out from society. A distinction is made between me and not-me. The text appears as an object and it works upon the subject because the subject articulates itself through the text. As Jay Bolter (1991, 210) noted, the reflexive character of writing gives the writer a new awareness of self. The writer observes her objectified or externalised other on the page or on the screen. The words stand out as if delivered by someone else to the writer, as if the writer was the addressee. Over time, it influences memory and gives continuity to meanings and attitudes. Writing creates a text and affects the author. It becomes an instrument for self-change.

DOI: 10.1057/9781137446466.0003

The paradox of the recurrent objectified text describes a modern fact of writing as a medium of identity-formation in modern societies and itself a hallmark of these societies. The truth is no longer out there, in old tales or from the reading of the Holy Book. Even the Bible now is more than anything else simply *a book*. For most people, Christians as well as non-Christians, the Bible must be read critically, interpreted, worked on, invested in and so on. This requires dedication and will. Basically, this is the case with all relations we have with those that George Herbert Mead called the Generalised Other, that are involved in our personal development. To be or not to be is a question of hard work. And this lends authority to the texts that we draw upon in our lives, authority not as the voice of tradition, or the voice of God, but as *produced* authority, authority as a written text.

Therefore, as any other Self, this Self has no final form. Unfinished selves that appear on the screen indicate the Self as something dynamic and unfulfilled. New life experiences are added, and old events are described with new words, with new programmes and revised design. In this way, this Alter ego Identity can be easily updated (and therefore also *outdated*). Still, a norm or a line of continuity must be followed, as it is the same individual who is presented in this running autobiography. What does the updating of the screen-based self imply for the subject? What kind of discourse is narrated? The blogger operates selectively. Cutting and pasting, writing, showing, linking, documenting, narrating, playing with words and images is the nature of autobiography, as it probably is for all sorts of self-documentation practices. For the reader, the personal homepage is not a miniaturised, textualised version of the author. It is *not* a projection, *not* an extension of the Self, it is a mask in front of someone, with a complex relationship to what is behind. The blog or web page as a writing tool for autobiography is designed for this shaping and structuring of a presented self. History and the past are easily wiped out, the mistakes and the disappointments of life disappear. The pure Self remains on the scene in an almost parodic position. It is a subject that acts upon the world, that makes achievements, that explores his world, that makes statements and that is conscious about his ability to change his living conditions. The blog therefore became a performative symbol, a writing act that gives the author a name in public or semi-public space in the universe of the web, and which places him or her in the world of meanings and people. Keeping a web page or a blog is to make space for oneself in a world of stories, citations and recitations. It is an act of

DOI: 10.1057/9781137446466.0003

identification, like showing a driving licence or a passport, only far more detailed and informal, and apparently digging much deeper in the Self.

In sociology, this paradox of freedom and dependence of social integration is theorised in terms of differentiation, system colonisation, trust and risk. Giddens conceived of trust as a medium of interaction with the abstract systems that both empty day-to-day life of its traditional content and set up globalising influences (Giddens 1991, 3). And what Habermas called 'colonisation' is a displacement of action from the hermeneutical processes of understanding embedded in lifeworld contexts which produces an instrumental and objectivating attitude, rather than an understanding of engagement with other subjects. In a system-theoretical approach, the development of a Self marks the manifestation of a personal autonomy from society, and at the same time the individual's increasing dependence on society. Society is characterised by individualisation and system-formation. Ulrich Beck and Zygmunt Bauman formulate this point similarly to Luhmann: society applies individualisation as a form, in which society can reproduce itself as society. Society cannot any longer integrate its members entirely through encompassing values and tradition – it has to find other ways, relying on communication networks. The credo is mobility and reflexivity under the condition of social contact!

Tertiary orality

The blog is produced by individuals (or families/private groups) in the mixed capacities of private persons, clients, citizens or consumers. The blog's relationship to its author is not one of truth, objectivity or accuracy. It constitutes itself as a relative autonomous space where the author invites himself/herself to a social space. The asynchronous nature of blog presentations makes them more comparable to personal and private textual forms like the notebook, the diary and the letter than to speech interaction. But the potential mass audience of blogs and personal web pages makes them into quite a different type of presentational medium. They are not only writing media, they are media for personal publishing, often composed as a multi-modal collection of texts and other media – and genres. As text, it is dynamic and flexible, some would say more 'alive'. It is far more synchronous than printed books. To some extent, it can be compared to a newspaper, which contains several media forms,

DOI: 10.1057/9781137446466.0003

and which is updated daily, within the same format and style. Both homepages and blogs are referred to as *personal journalism, self-publishing* and even *self-advertising*.

Speaking involves appropriation of language; it establishes a social bond with the other in a network of places and relations. These characteristics are also valid for practices like 'using' in day-to-day life, and create room for innovative practices about system preferences laid down in products. To indicate the boundaries of everyday life, Michel de Certeau (1984) made a distinction between *literacy/writing* and *orality*. *Writing* represented the formal power of documents, isolated from the context where it originated. This imposed an impersonal character on messages. In the scriptural economy, writing mediated strategic relations, whether scientific, bureaucratic or economic. It represented the modern way of defining lifestyle, morality and truth, and thus a way of codifying and controlling life. On the other hand, the *oral* exists for the speaking and the hearing, and is informal, emotional and intuitive in its nature.

de Certeau is probably right in that typed text is still the primary medium of modernity – TV and film never changed that fact. When the spoken word seeks authority, it needs the dis-enchanting verbal reflection in writing. This is the case in education, politics, economy and science. The word must distanciate itself from the simple and universal magic of the voice without any backup. As de Certeau writes: '... one can read above the portals of modernity such inscriptions as: Here, to work is to write, or: Here, only what is written is understood. Such is the internal law of that which has constituted itself as Western' (de Certeau 1984, 134). Orality became subject to a scriptural intervention and containment, as in TV and radio. This is Walter Ong's 'Secondary orality' (1982/1991). Even poetry had to give in to the typed book.

Today, we have entered another stage of orality. Now orality relies on rewriting, downloading, recording, transmission and networking. Oral voices are heard everywhere, but given authority through a number of presentation and storing media, from the blog to pod-cast. This is the 'third orality'. This new orality possesses some sort of autonomy, which gives it freedom to present itself in many different shapes. It is manipulated text in a much more subtle and complex way than the word. It is often involved in a strategic game. The page on the screen is a place involving not just utterances – it has become a more lasting product in time and space, which gives the text an objective and also personal touch, and therefore more responsibility. It not only *derives* from the external

DOI: 10.1057/9781137446466.0003

world, it turns *towards* the world from the outside and wants the world to conform to it. Unlike the spoken comment, the web page wants to be a social fact, if not in a big way, and regardless of the content as such. In this sense too, it is not only a statement, but also a perlocutionary speech act, an appeal or a demand.

The text consists of verbal text and photo and establishes a detailed and explicit system of elements, a complex and ordered composition or *bricolage*, such as the newspaper or a birthday invitation or a homepage on the web. The text may include several genres or media forms. Still, it is the verbal text that gives the other media elements a place in the hermeneutic patchwork and therefore the text as a whole is the shape of an ordered totality. Writing (verbal and pictorial) has primacy.

In this multimodal process of rhetorical convergence, both writing and images seem to receive new functions: As Bolter and Grusin (1999) argued, technology has the effect of making text image-like by representing its verbal structure graphically. On the web, the emphasis on design is radicalised. Java applets, icons, touch screens and clickable links give the writing a visual, symbolic style. This writing is more informal, inventive, less rule-bound and more rule-playing, more talk-like. Email and the blog seem to be another way to 'write the voice', as perhaps is the case with typed poetry. It seems like this hybridity gives this kind of writing a more oral style, and so opens the way to undermine the regime of the written. Maybe it gives the oral a new chance through these new forms of writing. Maybe this is a new way in which everyday talk makes use of writing to reach beyond then-and-there situations.

Nevertheless, to repeat, the semiotic difference between writing and orality prevails. They remain distinct systems of meaning-combating as well as supporting one another from without. Furthermore, the written document still has priority because systemic rationalities in science, politics, law, markets and education depend on it.

Talking/listening

While the telephone revolutionised talk as a collective medium for the working place and the family, the mobile telephone brought in the personal dimension. Technically it all started in the beginning of the 20th century, and socially, at the end of it. In 1910, the Swedish engineer Lars Magnus Erichsson built a telephone into his wife's automobile. In order

DOI: 10.1057/9781137446466.0003

to use the telephone, one had to stop the car and wire the telephone to the telephone wires on poles alongside the road. Power was generated by cranking a handle. This was the first Erichsson phone and a playful beginning of the substantial role of mobile telephony in Scandinavia (Agar 2003, 10). However, what made the mobile an 'impossible' medium up until the 1970s was not the idea of connecting telephones to one another through radio, but that the limited radio spectrum was already reserved for military and commercial interests. The requirement of one unique frequency for each phone call made radiotelephony a dream. Of course, the dream came closer to a reality when D. H. Ring developed the idea of cells that divided up the spectrum, which allowed for several callers to use the same frequency at the same time as long as they did not occupy the same cell. The idea was constructed in such a way that the system could identify all users, and the driver did not have to be concerned with passing from one cell to another. Still, even if Bell Labs invented the transistor that allowed for smaller and lightweight handsets in 1947, switching technology was too slow to handle the cell method. Unlike today, the technological development was almost completely the responsibility of large-scale national and quite hierarchical monopolies that did not rush to change the direction of a very successful ride.

Later, in the early 1990s, the US radio spectrums were auctioned or handed out through lottery in each city or district, which created a system of local monopolies and modest competition. Also, the receiver of a mobile phone call was partly charged for the call. This made it into an issue of money to keep one's number limited to some few callers. It was a kind of reversed network effect that served to keep the growth lower than in Europe. Also portable phones were not an immediate success because the car radio took care of much of the need in the car-culture of the United States (Agar 2003, 43). In contrast, in Scandinavia, the Nordic Mobile telephone system (NMT) and then the European GSM mobile system were products of long-term research hosted by the national telecoms. The radio spectrum used was considered a national resource that was dedicated to the new kind of phone. Similarly the new digital standard GSM in the 1980s and 1990s was made into one of the pillars of the European project; it was to provide a material basis for increasing cooperation in science and industry, and in the next instance allow stronger political and cultural integration. And in GSM, some small and hidden possibilities were waiting to be discovered, first of all SMS. The success was undisputed; the GSM phones were smaller, lighter and

DOI: 10.1057/9781137446466.0003

better designed. The first truly portable and personal phone had reached the market of personal communication.

In Japan, NTT first developed the first 'Shoulder phone', a car phone, in 1985 that could be carried around in a shoulder strap. In 1987, the first handheld cellular phone was offered by NTT, primarily for professional and organisational purposes (Okada 2006, 41). It was typically an organisation and not an individual that was listed as the rental subject, which indicated its collective and not yet personal status. Around 1996, the adoption rate for mobiles was 25 percent in Japan. But Personal Handyphones (PHSs) (see later) were still the preferred mobile device for teenagers. Partly due to the competition from the PHS, the telecom dropped its rental model on mobile phones and introduced discounts.

When the privatisation wave hit the telecoms in the 1980s (England) and 1990s, (Scandinavia) the infrastructure was all in place. In Japan, NTT DoCoMo launched a new digital standard in 1989, which created a fast growth of subscribers. DoCoMo's I-mode service took off around 2000. Like the WAP in Europe, the content providers of I-mode had to be approved by DoCoMo, but this meant quality as much as censorship. With the I-mode service, one could reserve tickets, check bank balances, transfer money, subscribe to a number of information services, and send and receive email. It became immensely popular among teenagers and young people in particular.

In the beginning of the1990s, *the pager* developed into a personal medium through a number of changes in Japan. In the late 1980s, and after deregulation of the Japanese telecom market, NTT marketed the 'Pocket Bell' pager, which showed digits and letters on a small display (Okada 2006, 43). The sender could type in a call-back number, subsequently independent of organisational limits. In conjunction with the phone, this made the pager into a useful device for private and personal use, particularly among teenagers and students. Students quickly relayed messages in their own codes, which we more famously know from SMS texting. This was largely a Japanese phenomenon. From around 1993, the clear majority of the subscribers were private subscribers. Typically the peak hour of use changed from early afternoon to 10 pm (Okada 2006, 44). In 1996, pager subscriptions hit around 15 percent nationwide in Japan. Among high school students, the adoption was close to 50 percent (Okada 45).

Soon, of course, the mobile took over. However, in Japan, a substantial part of younger users took a detour via Telepoint. London's Telepoint System was introduced in 1989, but was mainly in use in Japan, China and Thailand, and to a lesser extent in India, Indonesia, Vietnam, The Philippines, the United Arab Emirates, and Ethiopia (Kohiyama 2005, 65). Telepoint is a system that can either be seen as a digital element of the (analogue) home cordless extension of the regular telephone system, or as mobile telephony 'light'. Technically, it was a system of cordless phones connected to the fixed telephony system, and was generally conceived of as a mobile system. It consisted of several small antennas in an urban area to be used by small wireless terminals. In Japan, PHSs were launched in 1995 for Telepoint, and were far cheaper than the mobile phone, and the terminal was lighter, smaller and with a longer battery life. Telepoint could not be used within homes, but it was possible to call from some urban point and it could be maintained up to walking speed. However, already in the mid-1990s, it was clear that the operator could not make profits, partly due to network charges paid to the NTT. Terminal manufacturers lost interest, since the terminals were given away by the PHS providers. At the same time, the Japanese *keitai* (the Japanese mobile) approached the size, weight and battery life of the PHS (Kohiyama et al. 2006, 65).

As Okada (2006, 45) points out, the pager and the *keitai* began their career in professional spaces, and extended their territory as the flexibility of the media increased and users' demand and experimentation transformed them into private and personal devices. That personalisation involves de-professionalisation as the typical process of the telephone, pager, the mobile, the computer, email and the web. However, the process is even more fine-grained than the step from the office corridor to the living room and the pocket. First, most communication technologies emerge not in the business sector, but as military technologies, which subsequently get 'civilised'. This is the case for a wide range of media from the electro-magnetic telegraph to GPS. Second, as Japanese studies show (Yoshimi in Okada 2006, 45), the phone moved from the office, to the hall, then to the living room, then cordless and extension phones were set up in bedrooms, before eventually being further extended and transformed into the pocket phone.

To be sure, this transfusion from defence to business to the household and then to the individual is not without variations and exceptions. More importantly the journey will not stop with the personal terminal. We see a continuing personalisation of products and services on the Internet. Personalisation moves further – from individuals to things, from houses and cars to fridges, bikes and suitcases.

DOI: 10.1057/9781137446466.0003

Watching

Finally, a brief note on long-distance seeing. Around the turn of the century, mobiles quite rapidly became camera-phones, and less than a decade later, almost everyone carried cameras in their pocket. This phenomenon took advantage of another tradition, that of family snap-shots. Amateur photography reached a new personal and at the same time more public level with George Eastman's Kodak camera in the 1880s (Kitzmann 2004). It documented domestic events such as birthdays, holidays and weddings (Jacobs 1981/1986, Levine and Snyder 2006). The web camera wave probably got its boost from the Kodak, and later the one-time use camera was introduced in the 1990s, which made the family camera into an individual and more mobile device, and was very popular among Asian teenagers.

In Japan, Photo Club was another very popular service that preceded the camera-phone (Okada 2006, 58). Photo Club was photo-booths on arcades in the cities to take photos and make them into stickers. The 'cam-phone' quickly generated a variety of services on the Internet like iPhoto and other programs on computers, and sites like Flickr, Photo.no on the Internet. More importantly, during the first decade, third-generation mobile telephony was adopted considerably slower in the United States and Europe than expected by terminal and service providers. In Japan and South Korea, Internet on the mobile quickly became a success from the beginning of the century, and is primarily used for email to mobiles and laptops. This was a step forward, as SMS messages could only be sent to other subscribers of the same service provider. Other services currently provided to the mobile besides the Internet are location-based services (GPS) and TV. The introduction of Skype and competing services trans-formed it further into an even more universal machine for communica-tion. The mobile more and more resembles the internet-connected laptop functionally in that it is used for a variety of weak-tie purposes: checking one's business -mail, watching the news, trying to get in touch with the plumber, planning the week or ordering a haircut.

Individualisation

I have sketched some aspects of the transformation towards mediated personal communication, emphasising Asian developments in order to

DOI: 10.1057/9781137446466.0003

highlight cultural variation. My argument is that this development is deeply implicated in what sociology terms individualisation. Forms of individualisation or 'cultural privatisation' take place when institutional rituals in local communities, trade unions and churches have their influence and authority reduced as the private sphere becomes even more important, and when traditions and rituals take place in the home rather in the local institutions. This is interconnected with a growing emphasis on the individual as a meaningful unit for rights and responsibilities and leisure, with less confidence in politics and the news media, and with other local social changes such as immigration. Sources for social identification move from work and church to family, friends and the media. This trend of transmitting responsibilities from institutions to domestic space leaves the family with pressures that it does not have the capital (human, social, cultural) to handle, and therefore must delegate decisions to the individual. The individual must more than ever reflect and act on her own course of life as an individual in a social environment of strong and weak social ties. Individuals with the help of social resources, rather than groups, make decisions concerning the individual. This is also of course a question of social capital. Individualisation is thus often a side effect of de-collectivisation, which again is a consequence of higher standards of living, expansive capitalism, and deregulation and privatisation of public affairs. This does not indicate that social groups and neighbourhoods are devoid of social significance for the individual. Individualisation and cultural privatisation should not be exaggerated and taken for granted in social analysis. Communities of various forms tend to be important background-resources for the well-being of the individual. We are talking about a general trend, first addressed by classical sociology, and currently in, for example, lifestyle and consumption studies (Blokland 2003, Bell and Hollows 2005).

Personal identifications do not necessarily mean a sense of belonging and direct dependence, only familiarity and recognition, indirect interdependence and therefore sufficient trust. The lifeworld provides background-resources that enable the individual to make coherent interpretations of his/her social reality. Rather than only being members born into social groups like the family, they are, to a larger degree than some decades ago, also 'associate members' that gives them the freedom to select and make priorities among social contacts. Identification, as Blokland (2003, 210) notes, is a less straightforward affair, however still a social matter.

DOI: 10.1057/9781137446466.0003

Institutions are to a lesser degree 'social facts'. Communities unite, but they may also divide. Ties are of many kinds, involving the whole sociological repertoire of rationalities and group constellations. They create bonds and bridges. None of these possibilities should be *a priori* excluded in studies of everyday life. In the analytical landscape between rational action, personality theory and theory of morally integrated communities lie the pragmatic and critical observations of this book on person and his/her media.

Media of Self

In the early 1980s, Michel Foucault became interested in how individuals work to make themselves into subjects. There was a certain change in his interest from the discourses that objectify the Self, to the question of how the individual constitutes *himself* as subject. This became clear in the third volume of *History of Sexuality*, and in a number of his late, 'ethical' texts. Foucault addresses this as a question of how the individual forms himself/herself into an autonomous subject. The question for the individual, and more so in modernity than before, was: 'Who are we in our actuality? Who am I today – in the world I am living in?' The answer has increasingly indicated an autonomous subject. This reflection and the freedom and ability to act upon it puts the individual into a position where it can, with more or less success, influence itself by certain means that it has at its disposal. Today, we may say that a more independent image of a Self gradually appears as a model for self-development. Current ideas seem to indicate the process as follows: I am not just like others, I am something different from what the expectations say I am, I can be different according to situations, I can perform in different ways, *I can change myself!*

To understand oneself as a subject, the individual draws upon certain techniques, what Foucault called *Technologies of the Self*: 'Technologies of the Self [...] permit individuals to effect by their own means or with the help of others a certain number of operations on their own bodies and souls, thought, conduct, and way of being, so as to transform themselves in order to attain a certain state of happiness, purity, wisdom, perfection, or immortality' (Foucault, in *Technologies of the Self*). Foucault notes that the question of writing and the self must be posed in terms of the technical and material framework in which the Self exists. He traces such

DOI: 10.1057/9781137446466.0003

writing back to Seneca and Greek and Roman antiquity, where the ethos became to take care of oneself, and to know oneself. One had to occupy oneself with oneself. The primary technology of the self, of course, was writing. From Plato onwards, taking care of oneself became differentiated from concerns about political life. To be concerned with oneself became related to ongoing writing activity. The individual became a subject for writing.

The Notebook: first, the notebook or copybook (the hupomnemata) was used in Plato's time. It was used for accounting, public registers and as a personal notebook serving as an aid to memory. One wrote down quotes, extracts, examples, things that one had seen and experienced, and reflections. The notebook served as an objectified, materialised record of things of significance for the individual. This new technology, says Foucault in the early 1980s, was as disrupting to the ancient Greeks as the introduction of the computer is today. The notebook served as an object to be written into, to be constantly read and to be an object for conversation. It was always at hand in everyday life, and was important in the subjectivation of the discourse. The purpose was to capture what had happened, to collect the past, as a resource for the shaping of the Self. It was meant to counter distraction and sudden changes of mind, by directing one's attention to this gathering of thoughts and reflections. It involves a selection of different elements, and of omitting and adding. It was a personal exercise for oneself. The point is to constitute a sort of unity of all these fragments of experiences and thoughts.

The letter: the other early Technology of the Self is the letter. Writing letters involved that one read what one wrote and in this way also addressed oneself, as well as the one who received it. The letter did, unlike the notebook, constitute a way to present oneself to another. The writer becomes present to the one who receives it – both through the telling of the facts of his life, and through the assurances that the other is in one's thoughts, that they are together, in spite of the physical distance. In a sense, Foucault writes, the letter sets up a face-to-face meeting. And the letter sets up a gaze outside the writer. The letter looks back on the writer, and shows the writer for himself.

Later on, writing as a means for self-examination, and for the cure of souls, was and is important in Catholicism and later in the Protestant movement. In Christianity, this was developed further as question of morality and conscience. In this cultural context, to know yourself was, rather paradoxically, to renounce one's needs as in ascetism, as a

DOI: 10.1057/9781137446466.0003

condition for salvation. The confessional, purification and sexual abstinence are other self-forming techniques that belong to this tradition.

The 18th century was, as Habermas notes in his book on *The Public Sphere*, the century of the letter because of the more developed public postal systems in Europe, and literacy among the growing working class. The postal system enabled quicker circulation and more meaningful exchanges of information. The letter is a genre, involving a sub-genre such as the love letter. It unites the privacy of the diary with the confession. But it is also a medium, and the genre is coloured by the technical features of the letter (its private character, its storing and objectivation of the written word, its circulation in time and space, its rhetorical way of reconstituting the dialogue in spite of absence). In the 18th and 19th centuries, it developed into a medium for the family and private life, more distant from religious reflection than the diary. By the end of the 18th century, the literate population felt at ease with the new subjectivity it mediated.

In the 20th century, a number of other modern techniques of self-expression are late contributions to this long development of personal media, which serve as media of the Self. Current cases of personal media of self-presentation like the personal website and the blog, as well as a number of network-oriented sites, can be interpreted in this historical context, and so indicate what is new and not so new in this recent development. Drawing on and extending Foucault's work, we may bring historic light to some aspects of this technology that appears to be so fresh and new, but really falls into a long tradition of self-mastering and self-caring. To construct a personal web page is to equip oneself with a dynamic mirror where social relations and contacts, personal experiences, what one has read, seen and understood can be seen, not so different from Seneca's notebook.

Interestingly, Foucault discusses this in terms of ethics. The Care for one self is an ethical practice, because it is supposed to constitute the individual as a moral subject, a subject that is responsible for his own actions. For what is ethics, says Foucault, if not the conscious practice of freedom? Freedom is the ontological condition of ethics. Ethics is the considered form that freedom takes when it is informed by reflection (Foucault 1994, 284). To become a subject is ethical work. It is a self-generating ethics. Then the question is: what are the means by which we can change ourselves in order to become ethical subjects? In recent times, it has been viewed as an aspect of *Bildung* (education) in a broad sense. What are the instruments of ethics that enhance reflexivity?

DOI: 10.1057/9781137446466.0003

Media of self-presentation

Perhaps, Foucault exaggerated the self-caring argument. There is also another aspect, here, which Foucault and others could not consider in full: what is truly new with the web is the element of self-mastering through *self-publishing*. Now, not only The Other, or a few Others, enables the resonance for self-caring, but conceivably *the whole Internet world*. Even if the actual number of Others remains relatively few in most cases, this potential public recognition changes. Thus, we need to consider the vital technological differences that exist between the media that Foucault addressed and the web. It is quite interesting that this writing in public, this 'Tertiary Orality', is available for all, not only for intellectuals. Is it not the first time in history that the individual as a private person has the possibility to write whatever he wants to a public? Is it not the first time in history that an ordinary person can present himself in writing to an audience? The personal web page and related genres are the voice of the people, in a form that makes it visible and recordable; that is, *another* voice outside the edited reality of the mass media. This technology of the Self has become a publishing medium for the Self, a medium for public presentation of Self.

The web presents itself as a new medium of self-presentation and self-reproduction, along with a number of other more symbolic resources, if more general in their articulations. In other words, since this form of self-writing, this form of Technology of the Self, takes the shape of *Media of the Self*, the practice is not only turned inwards, and not only towards a distinct individual, as in letter correspondence. It is a form of self-orientation that takes a loop outwards to a larger audience. It lies in the technical nature of the medium. It is a presentational mode. In order to present oneself to others, one presents oneself to oneself. To write publicly is to write to oneself, in a different way than in the case with writing letters to another individual. It is to observe many others' observation of oneself, because one has to read what one writes. It is to place oneself in the centre of the public eye, even if the eye is usually more local than global.

This presentational mode leads to practices as *performances*. To understand such everyday performances, we must realise that everyday life is a drama. Dramaturgical action is one of the main forms of action models analysed in sociology. More than anyone else, this model is associated with, and elaborated by, Erving Goffman. Here, social action is understood as an encounter where participants appear as the audience for one another and not only act, but perform for each other. The performance makes it necessary for the individual to act in specific ways, in order to present himself positively, to get positive attention and to avoid conflicts and disappointments. The projection of a certain impression is done in accordance with principles of dramatic performance. Every social act has a performative dimension, a stylistic dimension, which is a confirmation of the social aspect of the act. We play, and play out different roles in a reflective way that may conform to the definition of the situation.

This is about how we all behave performatively towards the external world. This cuts across norms, instrumental aims, feelings and desires. It consists of this adopting an attitude towards the world and oneself in order to go on in everyday life, to keep what Goffman calls the 'Interaction order' in place. This is a dramaturgical model of interaction as a medium of self-presentation in social encounters.

A useful conception for understanding the difference between performance and preparation are Goffman's well-known terms 'front region' and 'back region', which refer to different motivations and conventions of conduct (Goffman 1959, Meyrowitz 1986). Goffman labels as 'front region' that part of the individual's performance which regularly functions in a general and fixed fashion to define the situation for those who observe the performance. 'Front region' is the facade; it is expressive equipment of a standard kind intentionally employed by the individual during the performance (Goffman 1959, 32). The front region is the place where the performance is given. On the other hand, the 'back region' is defined by Goffman as a place, relative to a given performance, with aspects of the activity, but consciously suppressed in the front region. In front regions, performers are in the presence of their 'audience'. In back regions, on the other hand, performers are sheltered from their audience and so behave differently, apparently more real or natural or relaxed, although back-region behaviour can be regarded as playing roles as well (Goffman 1959, Meyrowitz 1986, 30).

Goffman's metaphors aptly illustrate how individuals change behaviour according to context and how both kinds complement and support each other in reproducing personal and social life. Everyday life consists of a wide range of stages of all shapes, serving as back and front stages for each other in a mosaic of contexts, varying in size, temporality, explicitness, level of activity and hierarchy and so on. The web page is another very clear case of this, where the web presence is very carefully

DOI: 10.1057/9781137446466.0003

put together in a sheltered space, later to be observed on the front region of the screen. Goffman's work is about the nature of face-to-face encounters, and is a very suitable place to start in order to get a clear picture of differences and similarities between face-to-face interaction, and electronically mediated interaction.

Extended familiarity

After the depression in the 1930s, the conditions for mass consumption improved because of a stable increase in standards of living and public intervention in social security. Taylorism had already introduced principles for rational and efficient mass production. Fordist production of consumer goods accelerated after the Second World War focussing on the family as primarily a consumption unit. An emphasis was put on design, fashion, taste and improved ways of distribution and retailing. Middle-class norms achieved hegemony in consumption modes (Bell and Hollows 2005, 3). Through design and marketing, consumption goods achieved more explicit symbolic power and entered the expanding world of lifestyles more directly. A further differentiation of style and trends followed in the 1960s and 1970s led to deeper symbolic boundaries between generations. Production and marketing for the exploding consumer market headed a post-Fordist trend of the exploring segments and niches. The consumer was increasingly someone who made frequent aesthetic choices, to affirm cultural memberships and reproduce distinctions. These distinctions are connected to relationships to the labour market, as it is related to ethnicity, gender, generations and so on. However, increasingly consumption appeared as a series of personal and family decisions in a project of the Self. Lifestyle consumption is today investment in modes of sociability, with the individual as chief investor.

The path highlighted here can be addressed by a wide range of terms that characterise communication as consumption, with aspects of civility, friendship, intimacy and contact enrolled in commercial services and products. They involve relative trust and familiarity, referring to social relations where the communicating parts have sufficient information about the others to establish stable social relations. Talja Blokland (2003) argues that *the familiar* lies between anonymity, where information is minimal, and intimacy, which denotes relationships where the ratio of exchanged personal information is very high. Blokland

DOI: 10.1057/9781137446466.0003

(2003, 91) distinguishes between three forms of familiarity: private famil-iarity between people in private spaces, institutional familiarity between people who generally feel affinity to the same values and usually belong to the same category (a peer group, a school class, a football supporter club) and public familiarity between people who are anonymous to each other (as in a neighbourhood and in a virtual chat forum). Taken together, the familiar characterises the bulk of ordinary relationships in and outside personal media.

While communication based on the mobile and the computer is seen to serve loose social networks, it could be argued that it also serves to keep subcultures together. This is particularly the case for the mobile. While it clearly integrates youth culture and business cultures, it is also a channel for the individual who switches between cultures, from family to business, or from schoolmates to the soccer team. When people act as bridges between networks, the mobile increasingly is a prime medium. International data also suggest that the prime use of the mobile measured in frequency of calls is to call home, which could be as much a sign of family integration as of individualisation. Whether the increasing sensa-tion of necessity to communicate among family members is furthered by, or a consequence of the mundane mobile, the fact is that it is a tie to children and spouse. Another frequent form of use is to call friends and relatives, which similarly could be interpreted as practices leading to community and integration. What Habuchi (2006 in Ito et al. 2006) calls telecocoons are zones of intimacy where individuals can maintain their closest relationships even if they do not see each other much directly. This fact of course has much to do with establishing exclusive trust. Prøitz (2007) has studied such telecocoons through ongoing text-message conversations between close friends. The smartphone along with longer battery life, flexible pay-models and ubiquitous connectivity underline and facilitate loose couplings between individuals and their social groups and communities. Social integration slowly modifies itself; it gradually takes the form – not so much of networked individualism, but of *networked personalisation* (Wellman 1999, Wellman and Haythorn-twaite 2002). We are dealing with 'media of mass destruction'.

Although the owner of a mobile gets a rather 'personal' feeling about the device, it is not more of a body part than the wallet (see Habuchi 2006 in Ito et al. 2006). Certainly, particularly in its first years, the mobile was personalised through ring tones, background image and stickers. After the possession of one or two mobiles, the most significant

DOI: 10.1057/9781137446466.0003

personalisation steps are the implementation of the contact list and the selection of 'apps'. However, the mobile is included in a small number of objects that individuals constantly need to be aware of and bring, like keys, wallet, wristwatch, handbag and so on. Rather than embodied, these objects are personalised; they are made into natural and yet significant objects that are taken into consideration, especially before some form of mobility. Our relationship to the mobile is rather that of awareness. They are elements in the small-scale plans we produce and revise during the day. They are certainly personalised in several ways, but only to the extent that we are able to have a reflexive relationship to them whenever necessary. Precisely because our lives consist of plans and decisions, these objects also need to be kept at a certain mental distance, and thus be reached reflectively.

In these ways too, the person came into being as a modern subject; his/her senses were mediated and at the same time ensnared by convenience and progress. Identity and sociality benefitted from and grew dependent on media that not only explained the large world, but helped us reach the Other, and simultaneously the Self. Our lifeworld became rebuilt, not only in its human and social dimension, but in its very structure.

DOI: 10.1057/9781137446466.0003

3

A Networked Lifeworld

Abstract: *This chapter introduces 'everyday life' as a sociological category and examines more specifically and critically the concept of the lifeworld. It examines the significance of the concept from phenomenology to sociology, and points towards a revised notion of the lifeworld more compatible with a modern mediated everyday life. It links the concept of the lifeworld to the more recent Anglo-American concept of 'domestication', addressed more specifically as 'personalisation'. It connects everyday personal media use to Michel de Certeau's theory of strategy and tactics in everyday life.*

Keywords: communication theory; digital media; everyday life; lifeworld; media theory; personal media; social capital; sociology

Rasmussen, Terje. *Personal Media and Everyday Life: A Networked Lifeworld*. Basingstoke: Palgrave Macmillan, 2014. DOI: 10.1057/9781137446466.0004.

The classical sociologist Georg Simmel once wrote: 'On every day at every hour, such threads are spun, dropped, picked up again, replaced by others or woven together with them. Herein lie the interactions between atoms of society, accessible only to psychological microscopy, which support the entire tenacity and elasticity, the entire variety and uniformity of this so evident and yet so puzzling life of society' (Simmel 1997, 109). Here, Simmel comes close to describing what more than a hundred years later appeared as a networked lifeworld. In the following pages, I would like to examine the concept of the lifeworld as a portal to the meaning of personal media in their current use. In order to be useful for such empirical research, however, the concept needs to be revised on several points in light of tendencies in advanced societies in general, and recent media change in particular.

Everyday life

An a priori distinction between face-to-face contact and personal media use in everyday communication has little explanatory value and is increasingly misleading in the understanding of young people's everyday life. The distinction between face-to-face versus mediated 'virtual' contact is not of prime interest. Rather, the question is how individuals use their skills to maintain their everyday life with the networks and means they have at hand, along with the skills and reflections involved in reproducing their social ties. This would indicate how various media and social practices are interwoven, and may inform changes in everyday life, particularly among young people. Thus, the guiding distinction here is not direct versus mediated contact, but rather how sentiments of intimacy, trust and social capital are socially differentiated according to different social practices and spheres, with or without media. Rather than organising research according to a distinction left behind by most people, the question is simply how people handle their repertoire of media according to their needs, aspirations and actual possibilities. This is a *pragmatic* perspective on media use: people tend to choose among, and use media in ways that are practically (instrumentally) useful and convenient for them. Questions of media use are brought a step closer to a familiar sociological theme: how people get on in everyday life, related to socialisation, advancement, friendship, love and so on in a media-saturated network society.

DOI: 10.1057/9781137446466.0004

Social research has long since stated that materiality and technology are embedded in particular contexts and social structures that make up everyday life. We can only make analytical distinction like the ones by Venkatesh, Chen and Gonzales (in Kraut et al. 2006, 110). They distinguish between three main elements of domestic space: social space, technological space and physical space. Social space consists of members and activities, time spent on them and interactions in the home. The technological space refers to technologies and artefacts that are embedded in the physical space and used by the members in the social space. The physical space consists of the physical materiality and layout of the home. In real life, the three forms of space are interwoven.

Therefore personal media are always, if only partly, socially shaped in and through contexts of use and society. This is quite obvious for us now, but it surely makes it more complicated to establish general statements about social use of media in daily life. Additionally, with personal media, the materially based reaching out of the user, and the motivations and purposes, may vary even more than with the radio and TV. The structures that create regularities in the use of personal media probably belong more to general social structures of everyday life, its mechanisms of privatisation and intimacy, than to the technological features of the media.

What has been termed the 'transformation of intimacy' includes transformation towards mediated intimacy as a central component. The 'pure' relationship endures only through romantic love and cannot rely on underlying, sustaining support like paternalism and tradition and negative sanctions against divorce. As traditional expectations have weakened, it lives more on its internal communication in various forms. As such, it is a vulnerable and a risky journey. As we know, the dissolution rate is high. Communication, and therapy when it fails, is expected to carry the burden of integration of a relationship or a family for a long time period. This has, as has personalisation and individualisation, been interpreted as one strand of rationalisation or modernisation among many (Jamieson 1999, Giddens 1992). The transformation involved is intimate communication and mediated communication. Contemporary intimacy is, to a large extent, flows of words to which feelings and understandings are attached – in the text messages, the blog, the emails, the Skype and the dating website. This does not mean, as Gross (2005) notes, that traditions have lost their grip; what he calls 'meaning-constitutive traditions' continue to shape new intimate relationships. However, these

DOI: 10.1057/9781137446466.0004

are more to be located in communication itself and less in the values of society. This too puts personal media into a more strategic spot. Social identities are deeply enmeshed with individual identities – we build self-representations by linking to others (Baym 2010, 111). Online connections embed us into collective pools of information and social networks that invoke shared conception of insiders and outsiders. Social networks may confirm and intensify sociological variables like social status, social class and gender.

Everyday life is a mixture of the unnoticed and inconspicuous on the one hand and the partly strange and abstract arsenal of mass-produced goods and services for consumption on the other hand. Georg Simmel famously discusses this confrontation between the ordinary and the new urban life more than one hundred years ago. The familiar world of daily life is continuously in process, dealing with standardised and mass-produced objects and structurally planned environments. The media explosion of course is one particularly noticeable aspect of the last two decades. How does this disruption take place with the computer and the mobile placed in the terrain produced by the trivial *and* the abstract? Is it really a question of disruption or confrontation? In subtle ways, everyday processes adjust and absorb the news of objects and media, as they must with regard to news and information. This has interested researchers in ethnography and sociology for a long time, as they stress the appropriation of domestic consumption (Gershuny), objects (Miller), TV (Morley, Lull, Silverstone), mobile (Ling, Prøitz) and other media. Sociologists and theorists like Henri Lefebvre, de Certeau, Bourdieu and others have attended to studies of the ordinary with various motives, most importantly because life is lived right there – in the ordinary everyday life in fasions conditioned by class, gender, ethnicity and the brutal randomness of life.

The idea of the lifeworld is helpful here because it provides a foundation for a hermeneutic view on everyday media use as meaningful activities. Personal media use is fairly coherent and rational. Also, it may indicate how personal media transform everyday life, through the new forms of social interaction in time and space that media technologies provide. The question is how media technologies mediate and reproduce the lifeworld in different ways. The lifeworld concept, I argue, can potentially greatly help us to make sense of daily life changes if it can account for individualisation of identity formation and personalisation of media. An understanding of the particular values, practices, habits and rituals

DOI: 10.1057/9781137446466.0004

of domestic life is vital to the understanding and use of new media practices. As will become clear however, I think we need a quite different notion of lifeworld than what has become known through Habermas' theory of communicative action. In the following, I address the changes of everyday life in theoretical terms, and argue that we need another conception of the lifeworld to contextualise social interaction, with and without personal media.

Lifeworld in modernity

Generally, the idea of the 'lifeworld' is a key to the phenomenological critique of the philosophy of consciousness. While the philosophy of consciousness conceives of the individual as an independent subject vis-à-vis the world of objects, the lifeworld designates individual experience as produced by immediate interaction with the world. The unconscious, ongoing constitution of the lifeworld precedes ordinary interaction with the objective world.

Alfred Schutz (1967) converted central ideas from Husserl's phenomenological thought into sociology. He also derived his concept of action from Weber and viewed it as subjectively meaningful behaviour oriented towards the practices of other individuals. He did not, however, accept Weber's emphasis on purposive rationality, as he considered it a too narrow perspective on human life. For Schutz, the lifeworld is constructed through non-reflective practices, the natural, non-theoretical, taken-for-granted imperative of action in everyday life. The lifeworld is the world always within reach, consisting of significant projects within a definite time–space. The lifeworld is the *Umwelt*, the 'vivid presence' populated by fellow individuals, between past experiences (*Vorwelt*) and future projects (*Folgewelt*) (Schutz 1978, 136; Rasmussen 2000).

> It is central that first, Schutz understands the lifeworld to be an *a priori* dimension of reality recognised by common sense. Within this province of meaning, experiences are designated as unquestionable. Second, the lifeworld is an intersubjective, shared world. As the reality is self-evident to the subject, it also incorporates awareness about the experience of others. The subject takes for granted that this knowledge is accessible to others. Third, the lifeworld is precisely defined by the taken-for-grantedness. It forms settings where situational horizons shift, though it constitutes a totality of what is taken for granted. It is placed in the flow of experience as a given and familiar province.

DOI: 10.1057/9781137446466.0004

As we move outside the given setting, into the *Mitwelt*, new complexes of meaning open for us as lifeworlds. Schutz also distinguished between different zones of operation. The primary zone is the lifeworld in its real sense; a primary world within reach, with the ones with whom one has a sense of shared time. In contrast, a secondary zone is populated by contemporaries within a greater, and normally mediated whole, which may always potentially be included in the *Umwelt*. (Schutz 1978, 258; Rasmussen 2000)

For Schutz, the lifeworld stands in contrast to the increasingly anonymous and bureaucratic relations of modernity. He argued that the pressure on the everyday lifeworld from societal structures is a key problem of modernity. According to Habermas, however, the differentiation of structures in modernity and the less penetrating taken-for-grantedness in modernity cannot easily be accounted for. Since the lifeworld concept in Schutz is an *a priori* fact, it is unsuitable for investigation as to how it has changed historically, and how its social condition could be otherwise. According to Habermas, the problem with phenomenological sociology is that it constitutes a basis for not much more than descriptions of the internal occurrences in the lifeworld from the perspective of the members. Habermas thinks that the social phenomenology from Husserl and Schutz is unsatisfactorily connected to the philosophy of consciousness. It is too trapped in the internal self-interpretation of the members in the lifeworld, and is thereby inhibited in pinpointing structural problems.

Habermas thinks that the sociological turn in Schutz' contribution to phenomenological sociology is therefore incomplete and seeks to repair it through two important revisions. *First*, he adds a system perspective so that the lifeworld can be seen from without and from a developmental perspective. *Second*, from the lifeworld, communicative action develops as a *complement* to the lifeworld. These two moves make it possible for Habermas to demonstrate how the rationalisation of society creates structural gaps and conflicts, and opens up a potential for rational development – a fulfilment of the project of modernity.

However, in Habermas' theory of modern rationalisation, the concept of the lifeworld appears in different shapes and for different theoretical purposes. First of all, it is: 'a reservoir of taken-for-granted, unshaken convictions that participants in communication draw upon in cooperation processes of interpretation' (Habermas 1987, 124). The lifeworld is a source of situation-definitions that are presupposed by participants as unproblematic. It is an intersubjective world hermeneutically demarcated

DOI: 10.1057/9781137446466.0004

from the common objective world as well as from the individual subjective world and it is present as long as it constitutes a background for an actual context of action. (Habermas 1984, 70) As soon as the context itself is brought into the scene, it becomes a part of the situation and loses its triviality. It is then no longer taken for granted as an unproblematic background. The lifeworld '...supplies members with unproblematic, common, background convictions that are assumed to be guaranteed; it is from these that contexts for processes of reaching understanding get shaped, processes in which those involved use tried and true situation definitions or negotiate new ones... Every new situation appears in a lifeworld composed of a cultural stock of knowledge that is 'always already' familiar' (Habermas 1987, 125; Rasmussen 2000).

Practices are embedded in the unproblematised realm of the lifeworld: 'Communicative actors are always moving *within* the horizon of their lifeworld; they cannot step outside of it. As interpreters, they themselves belong to the lifeworld along with their speech acts, but they cannot refer to "something in the lifeworld" in the same way as they can to facts, norms or experiences' (Habermas 1987, 126). The lifeworld is, drawing on Parsons', differentiated into the structural components of *culture, society and personality*, each of which produces resources for maintenance of the lifeworld, which Habermas accordingly labels cultural reproduction, social integration and socialisation. The lifeworld reproduces itself through cultural tradition (culture) which supplies meaning, social integration through norms (society) which supplies solidarity and community, and through socialisation of the young (personality) which supplies 'I-strength', the competence needed to stand forward in relation to others. The structural components of culture, society and personality are realised linguistically in the lifeworld. They produce cultural understanding through symbols, regulation of action and personal identity. The lifeworld reproduces itself through all three levels when contexts are interpreted in relation to the social order culturally, socially and subjectively. If the reproduction of *meaning, solidarity or identity* fails, the lifeworld ends up in a pathological shape.

Linguistic interaction takes place against the horizon of the lifeworld, against an unproblematic background of convictions, derived from the interpretative work of preceding generations. The lifeworld, as a conservative counterweight, constitutes a complement to rational argumentation. The less a cultural stock of traditional knowledge is present, the more dependent the communicative situation is on rational

DOI: 10.1057/9781137446466.0004

agreement. The modern dilemma of the lifeworld is that it is located in the dynamic between the normatively ascribed agreement (e.g. tradition) on the one hand and rational, communicatively achieved understanding on the other hand.

However, Habermas' perspective poses problems (see McCarthy 1978, Bernstein 1985, Honneth and Joas 1991, Rasmussen 2000). Habermas initially introduced the concept of the lifeworld from the participant's perspective, but transformed it to an observer's perspective, as a concept for critical analysis. The concept then focuses on the sociologist as an observing participant, and becomes transformed into 'everyday life' (Schnädelbach 1991, 18). An unfortunate dualism appeared with this: 'The action-theoretic approach inserts at least two actors at the microstructural level of groups and leaves the macrostructural level to be modelled in terms of systems theory'. Habermas does not intend system and lifeworld to operate towards one another as macro- and microlevel (Habermas 1991, 262). Still, due to the connection of lifeworld to everyday life, and of systems to formal organisations and markets, this dualism not only appears as a methodological abstraction, but as a dualism that runs substantively throughout the theory (Krüger 1991, 153). What systems theory refers to in modern society, Habermas (1991, 256) argues, is more than analytical. Modern capitalism has created economic systems that operate as self-steering, functional sub-systems (Habermas 1991, 257).

This dualism in Habermas' theory is doubled with the dualistic notion of steering media (communication vis-à-vis money and formal power) as coordinating the lifeworld or social systems. The system-originating media (in contrast to communication) appear structurally as independent of or only strategically connected to action, containing a 'deworlding" effect, rather than *as media and outcome* of different action orientations. Habermas therefore has no satisfying solution to the vital problem of how coordinating symbolic media *transform* while they fulfil their mediating functions. A rupture appears between the individual action and the functional rationality of social systems. System and lifeworld tend to refer to different societal spheres (e.g. formal organisations vs. family life, the public sphere and social movements; see Habermas 1987, 309). This, I argue, prevents us from viewing lifeworld and system as *analytic* dimensions of society irrespective of domains, and makes it difficult to see digitalised processes of communication (within, as well as between systems and lifeworlds) as 'political', that is, as contested processes that reproduce both system and lifeworld aspects in distinct ways.

DOI: 10.1057/9781137446466.0004

Habermas distinction raises questions about the ontological status of the two categories: are we talking about social areas, action types or impersonal mechanisms? Are we dealing with a concept that can possibly be filled with empirical content in sociological research? Habermas distinguishes between lifeworld and system according to the following: *principles of co-ordination* (social vs. system integration through the media communication vs. money/power), *rationalities* (communicative vs. strategic and instrumental action orientations) and *spheres* (everyday life vs. economy, and public bureaucracy).

The status of the lifeworld as both a social–philosophical concept (differentiating between hermeneutics and systems theory) and a substantive sociological concept (differentiating between 'the world within reach' and formal systems) is confusing. The second notion leads to a sociological concept of everyday life, with which a practical understanding of mediated social interaction can be addressed. It differentiates between cultural zones on the grounds of the logic of day-to-day practices. However, if this second conception of the lifeworld seems most applicable empirically, it is problematic to call this perspective a lifeworld, since it tends to obscure the necessary distinction between phenomenological and sociological investigation. The first 'phenomenological' understanding of the lifeworld cannot be omitted, and it should not be mixed with sociological conceptions of social groups.

The agent cannot relate to the world without somehow being situated in the common world of fellow agents. The lifeworld is the subjective and experiental dimension of individual agency, constructed by intersubjectivity. In pragmatism, action is not the pursuit of ends that the agent establishes *a priori* and then resolves to accomplish (Honneth and Joas 1991). Communication technology is not mere material means at the disposal of human intentionality. Rather, there is interplay between intentions and impulses of the agent, and the possibilities of a given situation; we find our ends in the world, as they become reproduced and changed in the connection between possibilities and intentions. Only in this sense is action, whether it involves personal media or not, teleological and purposive.

The conflating of the lifeworld and everyday life, which seems to appear in Habermas' theory, leads to the reduction of the phenomenological argument, in favour of everyday life as leisure time, reproduction time, consumption and so on. This institutionalising of the lifeworld tends to ignore the fact that individual experience takes place wherever human beings act, regardless of institution. It confuses phenomeno-logical sociology with institutional sociology, transforming institutional boundaries into experiential horizons. The unhappy consequence is that the lifeworld is seen as something outside systems rather than as a pragmatic and experiential dimension of society in general.

The Lifeworld Today

Consequently, an idea of how such media may, through subtle herme-neutic processes, become transformed into *individual* lifeworld resources is necessary. The lifeworld-system model must be balanced with an understanding of the reversal process where lifeworld norms (traditional values, participation, compassion, morality) penetrate system media to the extent that they do not only serve system-integration. Examples of this may be social experimentation, political participation and coopera-tives on the market, hacker-cultures, social uses of TV and other mass media, and a number of practical, invisible processes in daily life. This reversal process is neglected in Habermas' theory. In an interview, for example, Habermas admitted: 'But this way of approaching systemically induced disturbances in communicatively rationalised lifeworlds was one-sided: it failed to utilise the whole range of potential contribution of the theory. The question as to which side imposes limitations has to be treated as an empirical question that cannot beforehand be decided on the analytical level in favour of the systems. ... colonisation of the lifeworld and the democratic control of dynamics of systems unrespon-sive to the "externalities" they produce represent two equally justified analytical perspectives. The one-sidedness of a view captured by a certain diagnosis of the time is certainly not inherent in the structure of the theory' (Habermas 1990b, 109).

The problems, routines and concerns of everyday life always revolve around generalising systems media in both constructive and destructive ways. Media of social systems are drawn upon as devices for manoeuvres within the modern existence of social systems. Everyday life takes place not only *against* the steering media of mass consumption, science and bureau-cratic power, but *through* them. They do, however, enhance individualised ways of handling problems and personal outlooks on the world.

A specification of the relationship between everyday life in the household and society with regard to media technologies is suggested by

DOI: 10.1057/9781137446466.0004

Roger Silverstone and associates. Silverstone characterises the household as a 'moral economy' (Silverstone et al. 1992). It is part of a transactional system of economic and social relations with the rest of society. The *domestic* communication environment consists of media with increasingly differentiated technical characteristics related to time, space and modes of mediating and presenting meaning. In order to understand the communication environment that participates in the mediation of the links between everyday life and society, their differences as well as their similarities should be elucidated: 'The technological culture of the household provides a framework for domestic, social and indeed political relations, mediating between members of the households and offering objectifications of their identities and competencies as well as mediating between them and the outside world. The domestic socio-technical system consists of a bundle of skills, tastes and competencies, expressed in styles and practices that construct and mark the cleavages of gender and age-based relations within and beyond the household' (Silverstone 1991, 141).

We should consider agency and its subjective and experiential dimension, the lifeworld, in relation to modern social systems. Then, the concept of the lifeworld does not belong to the sociology of institutions, but to the methodology of phenomenological sociology. The lifeworld is then understood as meaning-production, for the intersubjective, and yet individual, experiences and interpretation, the world and self-understanding in the different life phases of the agent. Further, the lifeworld is then connected both to the bodily *and* communicational spheres. When everyday experiences are mediated increasingly via personal media supplementing face-to-face encounters, transformation and extension of meaning from place-based to network-based lifeworlds occurs.

The 'lifeworld' is personal, tacit and reflexive considerations of personal life and integrity in the 're-embedding' of agency in the world of social systems. With the rise of digitally mastered social systems, the former lifeworld of unmediated social integration needs to be seen as differentiated, personalised and extended. The lifeworld appears individualised as the modern risk enhances a strong I-strength and ego-identity of creativity and competitiveness. Also erosion of the traditional community is compensated by modern support systems that provide security and safety (insurance, hospitals, police etc.), competence and information (education, the mass media etc.), spiritual support, work, culture (theatre, concerts etc.), capital (banks), food, clothes, housing and so on.

DOI: 10.1057/9781137446466.0004

The individual lifeworld refers to the personal politics of moral considerations of integrity, life plans and self-reflexivity as opposed to, and in confrontation with, systemic boundaries. The individual status can be characterised as both autonomy and dependence. Individual autonomy is reached through mobility in space and flexibility in time. Security is transferred to human-made, flexible and adaptive 'trust mechanisms'. Individual dependence is expressed through ignorance of the functionality of systems, lack of pre-modern trust mechanisms like religion and kinship, lack of contingency and as a general sensation of risk and insecurity. This is not to say that there is no life outside the social systems for the individual. Day-to-day life, in shorter spans of time, consists of unpredictable and coincidental encounters and interaction and so on. And yet, in the long run, the individual 'project' depends increasingly on social systems.

Domestication as personalisation

The concept of domestication is a good place to continue the tracking of what is going on, given that it too undergoes revisions on a number of accounts. In addition to media-related elements comes the complex social dynamics outside the media themselves that in various ways and for various groups domesticate technologies into particular contexts and culture. A way to start is simply to read the theory of domestication as a theory of personalisation. If we exchange the domestic for the personal, we may have a conceptual framework with which we may approach the personal 'taming' of various media, from the app-filled mobile to social media. Silverstone et al. (1992) distinguish between four elements or phases in the dynamics of media relations:

> *Appropriation* refers to the process where an object leaves the world of the commodity and is taken into possession by an individual. It becomes owned or possessed as cultural object and participates in a cultural totality. Acts of appropriation become a symbol of personal positioning and self-creation. Appropriation refers to both hardware and software, to the TV set and programmes, and to mobiles and apps.
>
> *Objectification* reveals itself in the display of objects, as they signal the person's place in the social world. It reveals itself in the usage and in the location of objects around the individual, how they are carried, ordered and arranged, with varying coherence and consistency, into the functional aesthetic of the person.

DOI: 10.1057/9781137446466.0004

Incorporation refers to the idea that media technologies are used in various ways, which sustain or change personal practices. Technologies in particular may be used for different purposes than intended by the market or in different ways than the usual due to specific personal needs and interests. They influence the usage of time, routines and rituals and so on. In this way, they may become taken-for-granted, or 'invisible'.

Conversion, together with appropriation, refers to the relationship between the individual and the world around him/her. Through the use of texts and technologies, the individual claims status in reference groups outside the home. The use of mobile on Facebook becomes a medium for achieving status and sociality, and a way to draw upon social and cultural norms of style, taste and morality as resources. This is the way fashion, TV programmes, computers and their software work, particularly for youth in relation to their peer groups. Conversion implies condensed private statements about the public culture and vice versa; they indicate membership into specific cultures by signalling judgement and competence of taste and style. Material objects both enable conversion and they are objects of conversion.

These are the ways that the boundary of the personal lifeworld is extended and blended into the public economy. The concepts are intended to provide a framework for the supply of meaning to objects by describing acquisition and use, in both private and public relations. As we see, they describe plausibly personalisation as much as the intended process of domestication. Although these four processes seem to refer to communication technologies as commodities and artefacts, and less as media for communication between the private and the public, they illustrate the process of appropriation noted above. The individual learns to apply principles of practice towards cultural products in new areas. The individual acquires the competence to reproduce certain cultural principles and to apply them pragmatically in new contexts. Habitus also inspires subjective probabilities that give aspirations and prospects towards social hierarchies. For instance, habitus provides the individual with classificatory schemes towards art and cultural artefacts. To a variable degree, artefacts have symbolic value. Habitus mediates differentiated appropriation of external facts. Accordingly, it provides systematically different interpretations of cultural artefacts (Rasmussen 2000).

The framework is clearly phenomenologically informed, and addresses technology use as interpretation and experience in the lifeworld. The non-instrumental aspects of the technologies, for example, their symbolic and cultural aspects, were emphasised as much as what they could do. This appears in the phases of domestication – and personalisation. While

DOI: 10.1057/9781137446466.0004

appropriation and conversion refer to particularly the symbolic aspects, objectification and incorporation refer more to the material expression. Briefly put, appropriation refers to the initial imaginations and projections of technologies as something that can fill a space or a need in one's life. The term objectification refers to the process of giving the artefact a concrete position. During incorporation, the technology finds its function or role in personal or embodied space. In the conversion phase, personal expectations place the images of technology in social space. This is a process of consumption as integration, of commodification of personal practices. It is about making things familiar and ordinary, and finding a place in a formal economy.

One important feature with several personal media is that one can be sure to reach a particular person, not only a family (home phone) or a group of chatting people. With media such as email and SMS, one can target particular persons, a fact that lowers the threshold for contact. Other people are not needed as intermediaries, and the significance of the medium as a trusted peer medium increases. A central theme in the following is that the marking of boundaries for communication is a fundamental aspect of human interaction, in spite of studies on the 'blurring of boundaries' or 'no sense of place'. Personal media are a modern marker for communication, often replacing the private/public boundary of the household. As Fahey (1995) argues, the public/private boundary in relation to the family may be simplifying a much more complex system of boundaries and zones of intimacy and privacy. Domestic space is undermined by personal (mediated) space. Livingstone (in Kraut et al. 2006, 128) argues that children, and even more teenagers, seek privacy and use online media to get it. The telephone provided ways to transcend the boundaries of domestic space for teenagers. With the text and voice functions of the mobile, the possibility to be in touch with friends late at night, when with parents on holidays and so on became more feasible. Studies indicate that most young people tend to use their mobile to call friends rather than using the home phone, despite higher personal cost (see, e.g., Ito and Okabe in Kraut 239).

Young people use personal media on the Internet to create boundaries around themselves and others, to communicate about embarrassing matters (sexual health), love and intimacy, and to experiment with identity and peer culture. What takes place is flexibilisation of the private conversation in the bedroom (with the sign Keep Out! Private! on the door). Now the possibility is larger in time and space to have such

DOI: 10.1057/9781137446466.0004

conversations through Facebook, however in a more rigid medium of communication than face-to-face talk. Personal media are actively used for privacy, for purposes of communicating in controlled circumstances. There are however boundaries that need to be marked and reproduced through frequent social interaction.

Everyday tactics

From Simmel, we are reminded that we must deal mentally with the dramatic changes involved in urbanity: noise, lights and speed of transportation systems, housing, schooling and shopping. From Walter Benjamin, we might bring with us the insight of the everyday objects that are not new and filled with the future, but the outdated ones, like the deep TV, the telephone in the hall, the stationary PC in the study. They are witnesses to how transitory norms about the convenient seem to be, particularly in an urban way of life. From Henri Lefebvre, we should keep in mind the point that everyday life is repetitious and habitual within 'constrained time and space', in contrast to the linear logic of markets and bureaucracies. We can pick up even more from Michel de Certeau (1984).

Michel de Certeau's perspective is more successful than many others in showing this linking of structural preferences of the political economy with the agency and construction of meaning, so that they are seen as a duality, that is, as both oppositional and conditioning. de Certeau (1984) examined the obscurity of everyday practices and its ways of operating things and articulated them into a sociology of everyday life. He suggested to: 'make explicit the systems of operational combination (les combinatoires d'opérations) which also compose a "culture", and to bring to light the models of action characteristic of users whose status as the dominated element in society (a status that does not mean that they are either passive or docile) is concealed by the euphemistic term "consumers". Everyday life invents itself by poaching in countless ways on the property of others' (de Certeau 1984, xii).

De Certeau focused on people's use of the media, on consumption of goods and services, on daily life in urban space and so on. People reproduce their everyday reality in 'hidden' ways, in different and scattered social areas that are defined and occupied by systems of production (TV, urban planning, telecoms, commerce etc.). These systems undertake a

DOI: 10.1057/9781137446466.0004

centralised and spectacular production that leaves no place for discourse on what people make or do with the products of the systems. People's consumption of products is on the contrary dispersed, everywhere, in a sense silent and 'invisible' because it does not draw on its own production, but on *ways of using* system products.

All these ways of using contribute to a cultural field that is 'unsigned, unreadable, unsymbolised' because of the pressure of the productivist economy, but which nevertheless is *there* in various degrees and patterns. Certainly, the practices are differentiated along socio-cultural lines. Similar strategic signals produce different effects. 'Hence the necessity of differentiating both the "actions" or "engagements" (in the military sense) that the system of products effects within the consumer grid, and the various kinds of room to manoeuvre left for consumers by the situations in which they exercise their "art" ' (de Certeau 1984, xvii). This cultural field articulates temporary conflicts and tensions, approximately culturally administered through symbolic compromises, balances and legitimacy. The forces of systems are met with the tactics of everyday consumption.

Technologies of power open a space, an interiority from which those elites may act on society (Feenberg 1991, 85). To act out its domination, systems require some space for operations. This space, however, cannot be reserved exclusively for the dominating part. Spaces are neutral only in that they offer themselves as sites for alternative manoeuvres as well. Consequently, the tactical operational autonomy of subjects corresponds to strategies. As opposed to strategies of power, *tactics* are the response of dominated subjects, their operating within the processes of the dominating system. Tactics are subversive rather than oppositional in that they operate on the terrain of the dominating, combine its elements into tactical 'tricks'. Tactic relates to strategies as speech (or rather, talk) relates to language. Tactics are the playing out within the game of other powers. They redefine and modify established values and practices and social forms. Andrew Feenberg calls this autonomy of alternative practices a 'margin of manoeuvre' (Feenberg 1991, 86). In everyday life, these manoeuvres take the form of informal communication, improvisations, unauthorised simplification of procedures, 'ignoring' and 'forgetting' orders, and numerous innovative modifications of established routines. Practices on the margin of manoeuvre may produce change (and thus be incorporated into dominating strategies) or they may contribute to the social order (like informal communication in formal bureaucracies).

DOI: 10.1057/9781137446466.0004

The concept of tactic suggests that deterministic perspectives on the technologising of the private sphere are unfortunate. From a 'macro' perspective, de Certeau's perspective negotiates between Habermas' reformulated critical theory of modernity and specific everyday practices. From a 'micro' perspective, it draws upon theories of tacit knowledge (practical consciousness, implicit knowledge etc.) and everyday practices in a world of social systems. Tactic is a form of practice that relates to the Other and the Other's space without taking it over, it plays itself out in sequences of time where opportunities are 'gently' seized. Tactics combine events and turn them into resources and opportunities. Most everyday practices are tactical in character, involving imagination, innovation, cunning manoeuvres, simulations, playing and discoveries.

De Certeau operates with a dichotomy that in some respects seems similar to Habermas' distinction between lifeworld and system. Everyday tactics of consumers in their lifeworlds stand against (although in a subtle fashion) strategic systems of production. In this symbolic battlefield, practices that can be empirically studied take place, which may refer to and further inform about the systems. In Habermas terms, de Certeau advocates empirical investigations of lifeworld practices, and their autonomy or interdependence with societal systems, whether political, technocratic, commercial, educational and so on. But in a much clearer way, de Certeau articulates the innovative and creative manoeuvres in daily life, playing upon structures whose system-oriented origin is alien to them. De Certeau explores incorporation and *appropriation*, the day-to-day operations that reproduce the meaning of everyday life in a productivist society. Because people, in their daily lives, are 'unrecognised producers, poets of their own acts, silent discoverers of their own paths in the jungle of functionalist rationality' (de Certeau 1984, xviii).

De Certeau's perspective helps us gain a better understanding of cultural activities as products of both systemic mechanisms and of the web of everyday life experiences. The transformation of systemic products by and through everyday interpretations becomes clearer. To disentangle this process, one cannot narrow the understanding of such practices to communication processes alone. One then leaves out the dialectic between the symbolic and the material, which is constitutive of daily life experiences as theme and context. The material and tech-

DOI: 10.1057/9781137446466.0004

nological reality provides a manifest fact to which people ascribe meaning and so incorporate it as rules and resources to their constitution of common sense. De Certeau emphasises that system media, as they are materialised as information, technology or goods, not only colonise everyday life, but also *become* colonised or appropriated by the dynamics and diversity of everyday life. Everyday life is a battle zone where rationalised and centralised system production is confronted with the quiet, differentiated and creative art of use in the trivial and invisible contexts of everyday life.

Roger Silverstone saw the relevance of de Certeau's perspective on the study of TV use. TV audiences do not read TV passively as a mass: 'We consume television not just in our relationship to the content of its transmissions, but also in our relationship to it as technology, as an object to be placed in our domestic environment and articulated into our private and public culture. Both sets of consumption practices involve us in a type of creative work. Our individual and social identities are defined through them. The paths we trace through television culture (the ways we talk about television, incorporate it into our gossip, or the ways in which we integrate television as technology into the pattern of our family life) are our own paths. We need to enquire into the specificities of those paths, their uniqueness and their universality, if we are to understand the dynamics of television's integration into everyday life' (Silverstone 1989, 81). The distinction goes between strategy and tactics, between agency and system. To Silverstone, de Certeau's perspective allowed for critical thinking about TV (or other media) in the mediation between everyday life and systems. It offers a framework for thinking constructively about media use as consumption, mediation and action. However, de Certeau exaggerates the ability in everyday contexts to converse and transform system output, which means that he is close to undermining his own critical perspective on strategies. Still, he presents an important approach to everyday consumption and uses of communication technology in several ways, which serves as an appropriate modification of Habermas' theory. Common-sense notions and practices in everyday life are not only a confrontation of the dominant strategies of systems. Everyday practices involve the linking of tradition, identity and system output. The meaning of everyday life is not constituted exclusively in opposition to strategies of systems; these strategies also provide the raw material for such construction of meaning.

DOI: 10.1057/9781137446466.0004

Relative distance

The main sociological connection between research on the use of personal media and general features of modernity goes through the notion of 'individualisation'. Individualisation is the way modernity has construed human identification since Kant and Rousseau. We teach our children how to make their own choices and take responsibility for their own actions. The child is to be given full opportunities to express its personality and individual difference, relatively independent of the historical and cultural past of family and relatives, which are now often seen as constraints for personal freedom. When our marriage does not work out as expected, we divorce. When we dislike the neighbourhood, we move elsewhere. After three years in a job, many look for new challenges. References to values of family and nation are not a priority in upbringing or in current pedagogical thinking. To be oneself is the true virtue, meaning that identity now is what is unique in one's personality. These virtues of difference and excellence in a democratic society open a space that needs to be filled by an emphasis on authenticity and self-construction of knowledge and norms. It is hardly necessary to enter the long debate from Kant to post-modernism on the nature of freedom to see that freedom is neither about complete independence, nor about cultural embeddedness.

On the one hand, freedom only rarely implies detachment from family and friends. On the other hand, the freedom we claim is a product of social interaction with others and with the past. As the later Foucault and many others point out, there seem to be two kinds of Self, the one that recognises its social foundations and another that suffers under the narcissistic delusion that freedom is private property, which one does not have to give, but only take. Diachronically, no one is the origin of one's self – we are all successors. Synchronically, we are all embedded in social contexts that are constitutive of our life and well-being. And for this reason, we do not have the power over our own life that we tend to imagine or are forced to imagine. Our life is conditioned by society, and the realisation of Self is a social construction. The Marxist-structuralist term 'relative autonomy' (referring to the State) could be used to describe the status of the modern individual vis-à-vis society. Another term is the late Roger Silverstone's 'proper distance', which in his conception refers to a balanced ethical approach between indifference and absorbtion.

DOI: 10.1057/9781137446466.0004

Let me illustrate by a brief note on Rich Ling's interesting study on the uses of mobile telephony in everyday life as forms of ritual interaction. Ling (2008) addresses mobile telephony in the light of the classic sociological theme from Emile Durkheim on social integration and social cohesion, or our sense of social solidarity. How is social solidarity generated in a modern and mobile society? Does the mobile telephone contribute to our sense of cohesion? Unlike Durkheim and Randall Collins, Ling argues that mediated interpersonal communication reproduces and develops relationships among those who are close. Ling argues that what he calls ritual interaction – in co-presence or mediated – is the glue that holds society together. The idea of classic sociology of religion is that rituals tend to enhance social solidarity. This may explain the universal and timelessness of rituals in functional terms. It is not difficult to see this in the mass media and broadcasting, but what then with mobile telephony?

The mobile is used for a wide variety of practical and social purposes: to make dates and appointments, to stay in touch and to coordinate everyday life. They may consist of very brief messages, jokes, gossip, shopping lists for the groceries and so on. Some of these conversations or text sequences may have ritual character, other may not. But they all expand and extend everyday interaction, Ling argues, and support the development of cohesion in this way. An important sociological point here is that social capital and social networks are forms of social interaction made extremely more efficient and flexible through modern personal media. The inter- and multi-personal in mediated social networks have become widespread precisely because of the multitude of innovations leading to the current forms of personal media. While Ling eminently demonstrates the role of the mobile phone for ritual interaction, the dramatic changes that this form of interaction enhances are somewhat underplayed. Ling rightly notes that such communication tends to be lightweight, simple, brief and often as a secondary engagement. It most often complements unmediated interaction among family members and friends. This points towards the observation that 'ritual interaction' through the mobile phone enhances a more flexible, loose and less constraining form of social solidarity. Social cohesion is probably not the adequate term for what the mobile phone generates sociologically in everyday life. Constant contact and interaction enhance elastic interaction and communication with less binding force. This does certainly not imply that the mobile phone makes personal or intimate

DOI: 10.1057/9781137446466.0004

relationships of less importance for the individual. It makes them more flexible. They encourage connections and networks rather than communities. To understand modern relationships and the role of mediated interaction to maintain them, we should look less on their normative or binding forces and more on how they make use of time in constructing patterns of contact.

In the *terrain vague* of everyday life, individualisation and digital personal media is a social realm of civility and solidarity, however usually without strong constraining values. The relationships are certainly moral, if rarely passionate. Intimacy, familiarity and friendliness may not only characterise atmospheres but also the proper distance that is needed. The point is nicely addressed by Goffman: our priority is to make everyday life work, and adjust the temperature of our social relations thereafter. What Goffman (1983) called *interaction orders* are reproduced – a substantive domain in its own right with its own structures and patterns. Interactions and conversation have an autonomous status insofar as they keep their own boundaries. Following Goffman (1967, 113), a conversation has a life of its own with its own heroes and villains. The features involved in a conversation between at least two persons create their own norms and ways, which make the conversation into the basic unit of a society. Interaction itself, not the involved individuals or their values and norms, is the unit of analysis, which everything else sociological is built upon. In fact, Goffman describes a social system similar to what Niklas Luhmann called interaction systems. Unlike Goffman, Luhmann considered communication as the basic unit, creating various forms of social systems (interaction systems, organisations, function systems and social movements). Goffman for his part describes conversations as a systemic element from which everything else social is built. Of course conversations take different shapes and forms, not only according to participants and immediate purposes, but also according to changing structural conditions. Personal media as new tools for personal networks influence Goffman's conversation in ways unimaginable for him and everyone else a few decades ago. A central task for a sociology of everyday life today is to enquire into how media create new and modify 'conversations' in daily life, and in the next instance the sociability of society, or to put it more in the line of both Goffman and Luhmann, how conversations reorganise themselves in order to make the new tools productive.

We may use the term *relative distance* to designate what I think is an apt sociological way of seeing the relationship between individuals, acknowledging of course that both isolation and embedding occur in different stages of life and in different cultures. *Relative distance* is both a descriptive and prescriptive term in that it somehow describes the current state of affairs in many aspects of daily life, *and* that it prescribes itself as a norm guiding the modern individual.

Personal media have become essential in 'identity politics' where relative distance plays an important part. Gerard Delanty (2003, 128) refers to 'personalism', to put emphasis not only on contemporary demands on self-fulfilment, but also on commitment, solidarity and collective responsibility. The self is seen as shaped through social participation and is sustained by a normative belief in collective goods. This, however, changes the perception of the social from a traditional lifeworld-approach to a looser network-oriented lifeworld. Rather than being integrated through given and strong values and norms, this kind of life-world is constructed by overlapping social relations. As Delanty (2003, 130) argues, they are products of practices rather than of structures. Or in structuration-theoretical terms, they are medium and outcome of practices. They are wilfully constitutions of both needs for belonging and independence.

In response to dark diagnosis of American civic life and claims about the decline of social capital in western societies, it is argued that the sociability of individuals simply takes new forms, less detectable by the community-oriented sociologist. If normatively integrated communities are losing ground in favour of looser social networks, a modified perspective is needed, which focuses on networks of association rather than on shared values and norms. If people are finding new ways to relate to one another, if they replace authoritative norms with communication and tasks, the focus should be on the associations, relationships and contacts, rather than on taken-for-granted values. 'Network integration' is a term that indicates these trends, which may help the sociologist to explore and explain the question of social order in the current society. That concept leads to a modified understanding of community as something deriving from social relations more than values, and that may be less stable.

Taken together, a 'thinner' or 'liquid' version of the lifeworld emphasises cognitive experience and information and weak-tie sociality, at the expense of immediate values of belonging, whether in spatial or cultural versions. A network conception of lifeworld is more flexible, more complex and more open than the traditional value conception. A lifeworld is today more reflexive in that its members are tacitly aware

DOI: 10.1057/9781137446466.0004

that it is embedded in an increasingly present and visual globality of overlapping communities and networks. To a large degree, communities are created through negotiated practices in order to enjoy a sense of belonging, and to avoid insecurity and loneliness.

A network conception of lifeworld may draw on network theory. That social communities can be seen as social networks is an insight greatly developed by Barry Wellman and his associates, based on the sociology of Mark Granowetter and others (Wellman et al. 1996, 1999; Granowetter 1973, 1983; Rasmussen 2007). Social network analysis does not take the normatively integrated community as a given, but is able to approach the question of lifeworld more subtly. It has led to the observations of non-local ties, adding other social variables for social organisation in urban settings, thus putting the often geographically defined community in a more nuanced picture. The emphasis on normative integration has been balanced by a perspective of social relations as complex, often non-local and not necessarily embedded in a community of values and norms. In much of Wellman's analysis, community is viewed as personal networks of a wide variety of tie types, and constitutes a valuable corrective to the lifeworld perspective. Generally, social network analysis has developed and refined an analytical distinction between sociology and geography, and thus helped sociology to avoid 'geographical determinism'.

However, analysis of social networks based on quantitative data tends to give the impression that personal networks are established instrumentally for accessing resources (Blokland 2003, 50). Community is often understood as simply interlocking personal networks. This leaves a lot unexplained, related to the stability and also the changes in social networks, which can only be explored by bringing in the wider social and historical context. Furthermore, a proper theoretical understanding of how people reproduce networks needs to be developed that matches findings of network analysis. Social network analysis tends to leave out too much of what social life is about, related to intersubjectivity and pragmatic, meaningful action. In a sense, the term 'networked individualism' is unfortunate because what we observe in contemporary urban settings is far from individualism, networked or not. To de Tocqueville, individualism meant love of family and near ones, but indifference to everyone else. Social relations are important but social and material and mediated contexts matter. If the community perspective overlooks sociability without normative integration, the personal network perspective may overlook memories and norms as well as traditions (Blokland 2003,

DOI: 10.1057/9781137446466.0004

60). The term leaves out too many sources of the social, and tends to ignore the fact that the individual is as social today as one hundred or five hundred years ago.

Individualisation implies some individual capacity for critical judgement and reflection, if social conditions allow it. As we know, observations indicate that much individualisation follows a pathological track for a substantial portion of the population in western societies, seen in the statistics of meaningless crime, depression and nervous illnesses, suicides, stress and borderline illnesses. Research on the social conditions of childhood in western countries reports on high degrees of harassment, lack of trust of parents, childhood pregnancy, school drop-out, discipline problems and so on. The obvious sociological diagnosis of this sorry state is that too many individuals are left alone with their self-development, or influenced in destructive ways by individuals and popular culture. Civil social networks are in decline in parts of society (Putnam 1995), which leaves a great burden on the shoulders of individuals who do not have the resources and support to manage 'self-realisation'.

Autonomy can only be reached if one is given the chance to acknowledge one's social peers and origin. Freedom can only be reached if interacting meaningfully with others. This diagnosis should not be confused with value-conservative complaints about the withering of traditional authorities and values. The point is simply sociological and political: the challenge is to help individualisation back on the constructive social track, leading to a democratic society capable of creating meaningful and relatively happy lives for most individuals and with public communication processes able to legitimate politics. The members of this kind of society must re-establish its bonds to ancestors and their achievements (not only their mistakes), as well as to future generations.

Analytically, a task for social research is to find a balanced way between individualism and exaggerated sentiments of social belonging, while leaving open possibilities for traces of both. Egocentrism and instrumental action are elements of everyday life, as is imagined and real communities. Individuals identify with other individuals, as well as with groups, and they constitute loose collectives. Research on personal media use in everyday life needs to be sensitive to a variety of sources for action on micro and meso levels. Our solidarities are based on both mechanic and organic constellations, leading to both Gemeinshaft- and Gesellshaft-like social forms. But mostly, everyday networks are hybrids of these, or to be found between them. The interest in sociology for

DOI: 10.1057/9781137446466.0004

analytical dichotomies has left a large, grey and seemingly insignificant territory unexplored between them, so far better addressed by literature and other kinds of fiction than social analysis.

The question here is not simply how people should be empowered through resources from society. Rather it is: do people themselves *already* have the means for re-establishing social ties and access resources they would need for a good life and a more just society? Could we trace such means in the very process of individualisation itself, not only in the established arsenal of welfare resources that are on the weakening front? Polemically asked: to sustain social networks, should we put our bets on the daily press or the mobile telephone? My point is neither that political nor cultural programmes to remedy social integration are mistaken, nor that they should be enforced. Rather, we should continue to look at the other end, by examining the social energies in the rapidly proliferating, digital personal media, media that are embedded in the wave of individualisation itself.

DOI: 10.1057/9781137446466.0004

4

Communication in Personal Media

Abstract: *This chapter introduces 'everyday life' as a sociological category and examines more specifically and critically the concept of the lifeworld. It examines the significance of the concept from phenomenology to sociology, and points towards a revised notion of the lifeworld more compatible with a modern mediated everyday life. It links the concept of the lifeworld to the more recent Anglo-American concept of 'domestication', addressed more specifically as 'personalisation'. It connects everyday personal media use to Michel de Certeau's theory of strategy and tactics in everyday life.*

Keywords: communication theory; digital media; everyday life; lifeworld; media theory; personal media; social capital; sociology

Rasmussen, Terje. *Personal Media and Everyday Life: A Networked Lifeworld.* Basingstoke: Palgrave Macmillan, 2014. DOI: 10.1057/9781137446466.0005.

In the 1980s and 1990s, research on the use of digital media was influenced by a North American fascination for virtual communities and digital identities. The emphasis was put on role-playing and experimenting identities as well as on electronic communities as new, democratic spaces on electronic bulletin board systems (BBS), MUDs and chatrooms (IRC). The theorising by early observers like Howard Rheingold (1993), Sherry Turkle (1997) and Steven Jones (1997) tended to stress the virtual as a separate existence, cut off from people's social status in actual life and everyday life circumstances in general. The unwarranted optimism probably derived from the novelty of electronic networks that were open and decentralised and emerged as a part of counter-cultures. It was also inspired by the writings of McLuhan and other media visionaries. A third source for this enthusiasm was post-modern speculations on the fundamental undermining of modern truth and certainty by popular culture (Poster, Jamieson, Lyotard).

In the late 1990s, this wave of irrealism waned as an effect of sociologically informed research on the relationship between online and offline practices. A number of studies on identity formation in MUDs found connections between online and offline existence, and that electronic social interaction was embedded in people's unmediated social life (Cherny 1999, Kendall 2002, Markham 1998). Generally, studies pointed out the connections between life online and offline. The reorientation came not only as an effect of sober analysis, but also as a consequence of actual media change. From the mid-1990s home pages appeared on the Internet as a channel of self-performance and self-presentation. From 1998 blogger-tools appeared on the net and from 1999, new blog and network sites like LiveJournal (1999), Blogger (1999), Wordpress (2003), MySpace (2003), Flickr and Facebook (2004) MSN LiveSpaces (2004), YouTube (2005), Twitter (2007). The sites are all dedicated to participation and user-generated content, which in systematic and accumulated form become added value for each user and create an exponential growth. The key is management of personal information and social relationships. Users present themselves in distinct and purposive ways that in accumulated form appear as social databases. Interaction of different sorts among users is stimulated and organised. There was no need for specialised computer or network competence on the part of the user.

In the 1990s, the main forms of online activity tended to be chatting, through emailing, chatrooms, the mobile and SMS and messaging services. Increasingly, software allowed young people to create content

DOI: 10.1057/9781137446466.0005

for web-based sites. The change was quick. Network development, improved user interfaces along with further individualisation, brought media change to the current stage of digital personal media. What were peak phenomena at the beginning of the millennium soon became mainstream. The affordability and improved user interfaces of multi-functional mobiles, laptops and mini-computers increased the competence and consciousness of both technologies and services on the net.

As digital personal media proliferated, pioneer users (the bases of early research) became the majority and young people with digital media experience brought their everyday life into new mediated forms. The new media were not simply alternative channels for interaction and expression, but media for the very social life they lived among friends, family and associates. For instance, Prøitz demonstrated how young people used mobile text messages and camera-phone images in their performance of sexuality and gender relations (Prøitz 2005, 2007). The use of mobile telephony that exploded from the mid-1990s was predominantly associated with maintaining existing social relationships. From the turn of the century, we have been observing a growing interdependency and infiltration between the online and the offline to the extent that the distinction has less explanatory value than in the 1990s.

In this and the following chapters I develop some suggestions for what communication and media theory we may bring to the workings of personal media in society. The topic here is communication theory for personal media. We cannot rely on theories of mass communications that were constructed to serve research on broadcasting and the press. I begin by briefly addressing the role of interpersonal in existing mass communication research to demonstrate this. But there is no need to start from scratch. Since personal media have inherited features from both interpersonal interaction and mediated communication, the trick is to carefully pick and develop insights from the existing reservoir of human research. In this chapter I therefore pay a visit to some insights from sociology and communication research to clarify personal mediated communication.

The interpersonal in the media

Interpersonal communication involves two persons or a small number of participants who exchange messages designed for those involved. The

DOI: 10.1057/9781137446466.0005

motivations, purposes are often shaped by the actual context, and it may have both instrumental and expressive aims. Broadly two lines of research on interpersonal communication are related to the mass media. First, there is the research on interpersonal communication in *audiences*. The interpersonal aspect has been present in mass communication research since the Second World War. It has been growing and can now be seen as an ever more significant aspect of the total communication context. An emergent mass media sector without direct contact with its audiences was the premise for the emerging research on mass communications and the mass media, and subsequently of theories of human beings as audiences and publics. The theories positioned people not as individuals, citizens, patients, students or clients, but as abstract members of uniform audiences. For instance, Lazarsfeld and his associates discovered (in *The People's Choice*) that mass-communicated political information was processed further, sociologically one might say, through interpersonal communication, managed by a particular communication role, the famous *opinion leader*. The receiving end was thus divided into two levels: 'leaders' and 'followers'. (They also applied the terms 'advisors' and 'advisees'.)

We also see this emphasis on the interpersonal in other macro studies of communication processes, like the agenda setting approach in diffusion studies, the spiral of silence thesis in reception studies and so on. In Katz and Lazarsfeld's study *Personal Influence*, the terms 'leaders' and 'followers' were applied as well, finding that leaders used the media more intensively than followers, and thus were more influenced by the media. In diffusion studies, notably in Everett Rogers's *Diffusion of Innovations*, six new communication roles, from innovators to lag-behinds, were conceived and tested on large-scale empirical material. Taken together, these studies, through the years, signified a more differentiated set of communication roles in mass communication, which not only transferred information, but also played a more normative role of influence and co-orientation (norm-building). The relationships between people, which applied the media as a reservoir of references and topics for face-to-face conversation, gradually received more attention in communication research.

Second, there are the studies of interpersonal communication, or simulations of such, in media *production* (such as studies of journalism and other genres in the press and in broadcasting). A number of mechanisms have been identified that serve as functional equivalents for a true and present communication partner. Phenomena such as liveness, authenticity, informality, proxemics and studio design all serve as simulations of

DOI: 10.1057/9781137446466.0005

communication, arising from the absence of interpersonal communication through the medium itself. Such media trends remind us that the interpersonal element has been present in mass media research since the 1940s, several decades before digital personal media appeared on the scene of private life. It has also constituted an increasingly important branch in studies of mass communication production.

Both in studies of consumption and production, the interpersonal element constitutes an added, contextual element, outside the actual transmission of the mass media. Interpersonal communication served as something mass media research imported in order to make sense of the contextual character of mass communication. Yet, whereas theories of effects gratification, influence, diffusion or reception lead to better understanding of the increased significance of the interpersonal in mediated communication, they cannot, since they are explicitly developed for mass communication, overcome its focus on interpersonal communication as something *external* to the actual work of the media.

Personal media then constitute a *third* connection – a synthesis – 'between' or beyond – the interpersonal and the mass media in that interpersonal communication is located right in the *medium itself*. This fact changes the communication situation dramatically, and in paradoxical ways. It provides its own (virtual) context, at the same time as the personalisation and mobility of personal media reduces the significance of the common material context as a reservoir of topics for communication. This suggests, among other things, that we need to study the media-internal context, as it appears as interface, genres, multi-modal composition, design and so on. Since the common social context for these kinds of interpersonal communication breaks down due to the fact that the communicating parties are distant from each other in space and time, the mediated communication itself provides some sort of virtual social context. The micro-oriented sociologist, let's say equipped with some of the insights from Erving Goffman's works, must look for contextual elements, or meta-elements in the mediated communication itself, which is coloured substantially by the nature of the medium.

The 'communicative turn'

It seems appropriate here to address the growing interest in communication as a compensation for overarching values and norms. The argument

DOI: 10.1057/9781137446466.0005

here is that this turn – called the 'communicative' turn – indicates the growing importance of mass media and personal media as generative for social integration. To throw some light on this, we may turn to some insights in sociology and social theory for the last 25 years or so. Along with the changes in media development I have sketched briefly earlier, new understandings of society have also emerged that underline the communication aspect of society. That is an expected effect of what the late Richard Rorty coined the linguistic turn in philosophy, with Ludwig Wittgenstein as a key reference. This turn implied that philosophical problems related to truth and knowledge as well as justice should be rephrased in terms of linguistics and language use. In all of the social sciences the linguistic turn has, along with technological change, left traces of a *communicative* turn, which emphasises the practical use of language in the construction of identity as well as in social inequalities and conflicts. Social action and organisations are seen as communication systems reproduced by observations and decisions. Society is seen as constituted in and through communication, which constantly reproduces and reorganises itself in order to cope.

This suggests that we need to address the process of communication and its media forms and genres, rather than (only) specific considerations of specific subjects. Let me explain this by briefly recapitulating some points in the process (the communicative turn) from subject philosophy to communication theory.

As part of his philosophical method, Descartes 'doubted' all existence, and through this could demonstrate that all wrong thoughts about the world were still thoughts about the world, which had to come from something other than the world. This was his way of proving the existence of God, which forced him to reconsider his doubt on everything. Husserl exchanged the Cartesian doubt with the phenomenological reduction, which begins with the fact that the world is as we experience it. We can become conscious about our consciousness about the world by suspending parts of it, bracketing it, to let consciousness affirm the existence of it. What then about the consciousness of fellow human beings? How do we meet the objection of solipsism?

To Husserl, intersubjectivity is the ability of subjects to overcome themselves, to communalise, where 'I' or the subject becomes a world for all. Unlike Descartes, this common world is not constituted through the relationship between consciousness and object, but between consciousness and phenomenon. The subject becomes integrated in a world, which

DOI: 10.1057/9781137446466.0005

is for others. The subject relates itself and understands itself through the interpretation of others who interpret the world from other points of view than its own. Objectivity was seen as intersubjectivity (Crossley 1996, 3). Through this interpretation and meaning-production, the subject becomes an intersubject. This process comes before the objective (scientific) perception of the object. From this one may say that Husserl was the last subjectivist and the first intersubjectivist.

However, Husserl did not take fully account of the role of language in the production of the social world. Husserl's ideas were elaborated in the social sciences by Alfred Schütz and the American pragmatists, the symbolic interactionists G. H. Mead and later Herbert Blumer, and in speech act theory. Jürgen Habermas was the one who brought the notion of intersubjectivity a step further by building on the linguistic turn as well as on Schütz, pragmatism and Goffman. As we have seen, Habermas projected the lifeworld as a mutually constituted taken-for-granted background context reproduced through socialisation, solidarity and culture. From the lifeworld, communicative action receives its competence to reason and criticise. Communicative rationality is considered precisely to be the grounded motivation to enter into rational interaction on issues of general interest. As reflexivity, it complements the lifeworld horizon, and still merges from it. The intersubjectively shared lifeworld becomes a sociological (empirical) reality, and communicative action a specifically modern effect of rationalisation (of systems, mediatisation and colonisation). Communicative action involves the automatic ability to produce some specific expectations of others through idealisations, that is, criticisable validity claims about the subjective, the objective and the intersubjective worlds as subtext of their arguments. Thus, reflexive communication grounded in cultural integration (the intersubjectivity of the lifeworld) generates rational intersubjectivity and mutual understanding – and possibly, consensus.

Habermas's concept of the lifeworld, as we have seen, rests on two empirical observations: First, it is an empirical fact closely associated with the concept of everyday life, although *how* the concept of the lifeworld is to be understood sociologically is not perfectly clear. Second, the concept of communicative action and communicative rationality pose the possibility of handling disagreements and conflict rationally. Conflict and disagreement are unavoidable in all societies, but only modern societies have the linguistic resources to transform them into arguments and mutual understanding.

DOI: 10.1057/9781137446466.0005

In contrast, Niklas Luhmann's theory of society has reserved a peculiar place for acting subjects. In complex societies, the individual has reached an independent and reflexive position vis-à-vis the norms of society, and is now considered to be outside communication and thus outside society. Autonomous individuals enable communication, and communication is normally *attributed* to individuals so as to make sense of it. The out-differentiated, self-referential communication systems (function systems and organisations and face-to-face interaction systems) operate as subjects. Communication is differentiated through semantic codes and various types of media.

Let me sum up the distinction between what we might call the norm-oriented perspective and the communication-theoretical perspective as follows: First, Goffman and Luhmann (unlike Husserl and Habermas) see meaning and norms (as the quintessential *social*) as *arising from communication*, rather than *producing* communication. Communication generates illusions of intersubjectivity, and it produces simulations of a common world. Second, to Luhmann and Goffman (again unlike Habermas and Husserl), the phenomenon of communication (neither the human subject, nor the text or work itself) is the crux of the social. This is the *Interaction Order* that Goffman quite famously proposed as a research theme of itself, or Foucault's discourse, or Luhmann's communication. Third, unlike the focus on community and values, this indicates that personal mediated communication increasingly presents itself as an interdisciplinary field of research, spanning the new and post-hermeneutic aesthetics of Martin Seel, Hans Ulrich Gumbrecht and Friedrich Kittler, to the non-normative, communication-oriented micro and macro sociologies of Goffman and Luhmann. In spite of the emphasis of much sociology on individualisation and the focus on particular texts, such as films, advertising and computer games, both the human subject and the singular text have less explanatory value compared to mediated communication as the generator of sociality. Correspondingly, a readjustment (or widening) of focus in media research is needed (and underway).

Luhmann on communication

Interpersonal communication in personal media changes the significance of media in communication models. Consequently, other

DOI: 10.1057/9781137446466.0005

communication theories and modes appear more relevant. One such relevant theory is Niklas Luhmann's 'subject-less' communication theory already mentioned, constructed in connection with his sociological systems theory. Luhmann's theory accounts for construction rather than transmission, since the sender does not give up anything that the receiver acquires. Furthermore, communication is not about the ability of the sender to get his message across, but about the understanding of the receiver, and what he or she does with it. The identity of a message lies in the *reception* of the message. For Luhmann, communication is a result of three selections: *information*, *message* (Mitteilung) and *understanding* (Verstehen). Communication only constitutes itself as a fact if all three selections are made. Information is the selection of something in the world that can be uttered in some way. It is the meaning-construction of something that potentially can serve as a message. It actualises something from the infinite latency of human existence.

Next, a message must be selected from the information and formed as a message. This refers to the expression of the information in some distinct fashion, through some language or image, through the telephone or TV and so on. Information must be given comprehensible form. An utterance is interpretable as a selection. Only when a message is produced does the third selection – understanding – come into play. Understanding refers to the change of the state of the one who receives the utterance. Understanding is a selection based upon a distinction between information and message. However, for more communication to follow, the receiver must confirm or manifest him or herself through some reaction to the information or message of the sender. Therefore, there is often something in communication that works to enable new communication (through questions, friendliness, counter-arguments, provocations, appeals, etc.). Social systems can only reproduce themselves if communication follows communication. This is, I may add, precisely the constant challenge for communication in modern societies. It cannot rely fully on moral values, as was more the case in traditional society.

A dimension of the success of personal media lies in the combination of their motivational and pragmatic character, in that they carry a systemic rationality as both aims and means. The pragmatic aspect lies in their practicality and convenience, as a resource at hand to reach others, retrieve information, be entertained and so on. Personal media are resources that we use to act upon information and meaning from

DOI: 10.1057/9781137446466.0005

others, because such action usually appears easy and pleasurable. The threshold is lower than the gratification. This also entails that they are successful in narrowing our latent aims of action by presenting a distinct set of possible ways to act. As practical and useful means, they influence our motives and goals for action. They are thus not only resources but also motivators in that they quickly and conveniently present the other person or persons the useful information in and through the medium itself. To be sure, when we order a book at Amazon, the book arrives in the post some days or weeks later. Online ordering is a convenient instrumental action only. The postal system and the book medium must take care of the rest. In most cases however, as when we call someone, or respond to someone's posting on Facebook, the social contact, the social purpose, the meeting, is immediately established. The laptop and the mobile now present a fairly broad repertoire of output and input channels to persons, organisations, data bases and so on, which give us access to both, and actually present, people and information, in one and the same operation. We do not really think of the cultural and social possibilities as narrowed at all. On the contrary, we are all amazed about the new digital possibilities. Precisely this invisible selection of communication possibilities is at the heart of their success.

Karl Bühler's communication theory, on which Luhmann builds, was published in the 1930s and inspired the theory of speech acts by Austin and Searle where distinctions were made according to what is emphasised in the speech act. In Bühler and later Luhmann, speech acts and their intentions are not really relevant, only the possibilities inherent in language to fix information in an information universe, to utter messages and to allow for understanding. The three selections are not separate elements but components of a unity of communication. Communication happens when information, utterance and understanding are achieved (Luhmann 2012, 216).

Information is always a part of communication and always operates within a system. A sentence, a document, a statement, a lecture and a TV program are all chosen from a set of potential sentences and so on that could have been presented. What is actually presented depends on what has been presented beforehand in a conversation, in a research discipline, in TV schedules and so on. This background of information constitutes a context for the one who receives the message, which allows for such messages to occur. To follow a lecture or see a TV talk show, it is necessary to know something about previous lectures or talk shows in

DOI: 10.1057/9781137446466.0005

order to make sense of such utterances. Information is information about the selection of possible understandings, and assists the understanding as such. Thus, receiving the message or utterance in a mere technical or perceptive sense is not enough; the risk of not understanding or misunderstanding would be extremely high. The utterance is a coupling to or a selection from information, which is observed by someone who has the task of understanding. The lecturer or the TV program presents specific messages against the background of the information it has selected itself from.

One could certainly discuss what is actually the utterance in a phone chat, a lecture or a TV program, but it is not necessary here. The point is that it helps to carry communication further by producing a distinction between itself and information, making it possible for someone to observe, understand and possibly respond to with another distinction between information and utterance. The understanding concludes the communication and possibly begins another communication sequence. If understanding does not occur, the student leaves the auditorium or chooses to quit the course; if the TV watcher changes the channel and so on, the communication ends. Still, group or mass communication may proceed as long as there are others who enjoy the utterances. In personal communication with or without personal media however, the communication ends. Note that misunderstanding and non-understanding may serve as communication. The student may ask questions to the lecturer who is then given the chance to specify her explanations. The receiver of an invoice may complain to the sender. The TV station may, when ratings decrease, take a show off the schedule.

Information, utterance and understanding constitute a unity, but among the three understanding is the truly critical component that sees the distinction between information and utterance in order to continue with more communication (Luhmann 2012, 220). Understanding must relate to the utterance-in-context. Unity happens in understanding, including misunderstanding and disagreement. Such responses also often tend to provoke response (what did you mean? I disagree!). However often this does not happen in part because it may not be technically possible, as when watching a TV program or when the student is too shy to ask questions.

The speaker will attempt to anticipate his or her audience or conversation partner, to estimate what is needed to entertain, engage, or perhaps to provoke in order to establish a conversation, an intimate relationship,

DOI: 10.1057/9781137446466.0005

to attract an audience or in short to create a circularity of communication. Rhetoric and conventions of genres are helpful here, and so are disciplinary concepts, humoristic clichés and other language-immanent resources. If however understanding does not occur, what happens next is entirely open.

The information concerns what we are talking about; the utterance concerns how a specific selection of this is communicated (written, shouted, broadcasted, etc.) and understanding concerns the registering of the utterance qua utterance. A unity appears from the three selections or components that may have feedback effects on those who make communication possible and who may continue the communication.

Luhmann's communication theory begins with the face-to-face communication where simultaneity and mutual physical context and immediate perception tend to reduce contingency; that is narrow down the range of possible utterances and understandings. In mass communication, strong and stable genre conventions, familiar personalities and external emotive means like music and studio applause help the communicating sequence to go on. In analogue writing (script, letters, books, etc.) the writer is alone and can only anticipate or imagine his or her receiver or audience. From this fact, a history of linguistic, semantic and literary innovation – most of all rhetoric – has arisen. The aim is seduction: to make communication continue by creating understanding among the absent in time and space.

In written communication through electromagnetic telegraph, email, blogs, text messages and social media, there is normally no common context in space or time even if the possible audience is narrowed down considerably. One has to rely on imaginary, motivational aspects in language, perhaps supported with images and design. In mobile telephony, the time frame and the oratory context enabled through a familiar voice is mutual. The geographies are however uncertain and often changing during the conversation, and provide little help for creating unity of information, utterance and understanding. How is communication possible under these new conditions? As we know, misunderstandings are frequent in social media and ignorance is as well. All the same communication continues. Blogs and Facebook and other similar platforms for communication have developed yet another set of technological features in addition to the ones inherent in language, such as specialisation and motivational aspects. These I will return to in another chapter.

DOI: 10.1057/9781137446466.0005

To Luhmann, the anticipation in utterances to seek understanding on the one hand, and the contingency present in all communication on the other, can be grasped with what he calls the theory of bifurcation (Luhmann 2012, 228). For communication to go on, for communication to be *anslußfähig*, responses with the principal 'yes' and 'no' ought to occur with equal frequency in the long run. As mentioned, misunderstandings and non-understandings are often occurring, but cannot occur all the time. The bifurcation – the choice between yes and no – takes place in every communication, and often the no enables communication to continue. Disagreement leads to justifications and new disagreement and so on, perhaps to the point where everyone involved agrees on their disagreements.

The observer who is to understand the utterance must understand what is uttered in relation to the world it is selected from. Or: the observer must ask: What distinction is operating here? Then the complexity remains reduced – the world of possible communication is contingent and rests on the distinction in use. This is second-order observation that dispenses with ontological truths, and as a European specialty, seeks what lies behind. We are more and more aware of the fact that things may not be the way they seem and that they may be different to another person. What is it that the other does not see? And does the other see that he cannot see? None of us have the complete picture. As Luhmann formulates it: We may be dealing with a circulation of a blind spot (Luhmann 2012, 115).

Luhmann has suggested (from C. J. Friedrich) that authority may be understood as 'the capacity for reason elaboration' as the ability to give reasons (Luhmann 2012, 224–225). Authority is in the one who always asks what reasons one has and finds further reasons for these justifications; this functions as an absorption of uncertainty, which since Herbert Simon's work in the 1950s has been considered as a critical function in organisations. Unlike Habermas, Luhmann sees this ability not as practical reason, but as a form of power capable of reducing complexity and defining topics on the ground of reasons. This authority is in fact necessary for communication to go on: 'It serves as the precondition of continued autopoiesis [self steering] and the precondition of connecting operations' (Luhmann 2012, 225).

Communication happens when a piece of communication is passed on as an utterance or message for someone to understand. The operation of uttering simultaneously creates a distinction between hetero-reference

DOI: 10.1057/9781137446466.0005

(information) and self-reference (utterance). Re-entry takes place through the copying of something into itself by itself. Communication creates or reproduces a social system and is an internal operation.

In personal media, one does select not only understandings, but also information and messages. In a Luhmann's frame of reference, one may argue that mass media require reception, which entails selection of understanding, whereas personal media require selection of information (hence co-production), messages and understanding. In contrast to mass media (and intermediary forms of interactive media) personal media are involved in all three selections involved in communication sequences: selections of information, messages and understandings. In other words, communication in and through personal media draws upon selections involved in both mass media and personal communication. This also gives a communication-theoretical account for why personal media are not well suited for particular sectors, or in Luhmann: function systems. Unlike mass media, which have out-differentiated their own function system, personal media do not mediate any specific set of communication codes. Personal media are as such suited for *all* social fields and systems and organisations in society, including private, interpersonal communication.

Luhmann (1996) argues that in the modern society of functional sub-systems, one particular system has been given the task of selecting information for societal interaction among social systems and individuals. The media system produces communication about communication produced by other social systems, including the media system itself. Information and debates about unemployment, financial crisis, elections, the world cup, scandals, the incompetence of teachers and so on are all second-order communication, produced by the self-reference of the media system.

But media forms riding on the Internet do *not* fit within the demarcation criteria of the mass media system. True, email, web and social media certainly diffuse information across time and space to millions of people. Furthermore, news, advertising and entertainment are to be found on the net. Still, the web was transformed through protocols of interactivity and co-creation (Web 2.0) into a communication medium. The Internet cannot be considered as any other kind of dissemination media in the mass media system or any other particular social system. On the contrary, the rapid diffusion of the Internet in the 1990s should partly be explained with its *complementarity* vis-à-vis the electronic mass media, as

DOI: 10.1057/9781137446466.0005

an infrastructure for personal media and interpersonal and networked communication. Communication on the Internet is not restricted to the mass-mediated information codes, which shape the principles of news, advertising and entertainment. The openness of the Internet makes itself useful for all sectors of society. As the mass media present their messages through the code of information/not information, this fact restricts the mass media from participating in *other* function systems and their internal information requirements. Although people in the financial business read the yellow press, they cannot rely on the press for internal, detailed, restricted and personal information. Scientists cannot publish in popular science magazines only, and the political system cannot use the mass media for their internal exchange of information.

In contrast to mass media, personal media normally produce groups and networks rather than *audiences*. An audience is the receiving environment for mass media content. It has no concrete status and does not consist of any particular number of individuals. In mass media, audiences are rarely spatially located. This abstract character of the audience means that it cannot be identified as such. It can only be observed indirectly. 'The audience' represents a range of *implied* or assumed ways to receive mass media output, an image or construction from the side of media organisations. It signifies a host of anticipated and assumed schemes or codified practices, enabled by the joint products of mass media output and intelligible individual media habits. Therefore, the mass media can only gradually improve and adjust their media output according to attention responses and indirect feedback with the help of the ratings industry and through reviews and debate in the public sphere. In other words, audiences do not consist of individuals but of measurable aggregations of typifications, images or models of media-use behaviour. Only in this way can the audience be transformed into a political community – or a commodity.

In personal media, including Twitter and Facebook, audiences are not invisible, they do not exist. In their absence, a number of individuals are performing communication in groups and *social networks*. There is no operative separation between production and consumption, and consequently readers are always potential producers of messages. Furthermore, personal media are available in and for *all* social systems, not only for what Luhmann calls the mass media function system. The rapid diffusion of the Internet in the last decades is little different from the dissemination of analogue telecommunication from the latter part

DOI: 10.1057/9781137446466.0005

of the nineteenth century. Its flexibility makes it applicable for military, governmental, economic, private and probably all other communication processes in society. It encourages rather than prohibits interaction. The information is produced by the communicating parts (as in email) or by small-scale publishers (as on the web) rather than by a gigantic media industry only. The Internet refuses to be restricted to a singular social system, it rejects the favouring of one particular code, because it is to be allied neither with a symbolic medium like money or power, nor is it a dissemination media of the non-interactive profession-based sort.

A simple sociological distinction throwing light on the difference between the genuine features of the Internet and traditional mass media is between category and network, stated by Harrison White (1965/2008, see Calhoun 1995, 220). *Category* distinguishes a group of people by their common features. It refers to boundaries, which the individuals in a group have in common in contrast to the external world, whether geographic, cultural or otherwise. Category defines the basic conditions for membership in a certain group or movement. It characterises the similarities of the members in a group which distinguish it from other groups. Cases, on the one hand, are members of a local or national community, subcultures with some common characteristics and so on. *Networks*, on the other hand, refer to a particular group of individuals among a social whole delimited by the social relations *between* the individuals. The network aspect refers to the density, durability and multiplicity of social ties irrespective of common features, geography and place.

We can connect this simple distinction to media technologies in order to see the features of different mediation processes and mediated integration. Television and radio establish cultural communities by exposing their common features to others. Mass media enhance social change by visualising living conditions for certain social groups who then become defined as a *category*, like the poor, students, immigrants, Roma people, gays, members of a particular community and so on. Television, in particular, visualises common characteristics of certain individuals and as such defines them or confirms the definition of them as a social group. The reach of TV is global, irrespective of age, gender, class or nationality. It has largely transcended class distinctions, even if what people actually watch differs. For technological and other reasons, television has, in the post-war era, become the all-inclusive medium that reaches everyone, occasionally with the same programme at the same time.

DOI: 10.1057/9781137446466.0005

Somewhat crudely, we may say that while the mass media meditate contexts as *categories*, personal media forms mediate contexts as *networks*. Compared to the category making of the mass media, personal media based on the Internet and telephony enhance *networks*. A set of email sessions does not expose or demonstrate in public common features or similar characteristics that would construct them socially as a group. Rather, they define groups through their actual mediating of social relationships. The characteristics of the group would appear through the variability, durability, regularity and density of the interaction. However, personal media like social networking sites and blogs construct hybrid forms. Boyd refers to networked publics as reconfigured publics with network technologies. What Chambers (2013, 57–59) calls personalised networked publics, particularly in social media, highlight personal control, 'yet also the problem or challenge of the nature of the interaction'. In the context of social media, 'listed "Friends" become personal "publics"'.

Plurality of communication forms

On the one hand, unmediated face-to-face interaction is probably considered to have a prominent, ideal-typical status among most people. On the other hand, there is no reason to consider the unmediated as a pure or ideal form, isolated from the diversity of media use. Mediated interactions are as real as unmediated communication – both in terms of interacting persons and topics for communication and for friend-ship and relationship to develop (Tanis and Postmes 2003, Brandtzæg and Stav 2004). The mediated and the unmediated are not different ontologies, nor are they different social spheres. Various media influence communication and interaction in different ways, but they all support the individual in his and her everyday interaction and meaning-production.

Social motivation is the main reason for participation in social network sites (Brandtzæg and Heim 2007). Blogs, for instance, appear as individual and even intimate forms of self-expression through text, photos and videos, but are also a way of maintaining social interaction with others. Users create profiles dedicated to online and offline friends and visualising networks of relationships, which all have real social effects.

As we are well into the second decade of the new millennium, new technologies for perception, interaction, publication and archiving are

DOI: 10.1057/9781137446466.0005

constantly thrown at the consumers. Many of them have quickly been absorbed by, and embedded in everyday practices. To most people in countries far beyond the developed part of the world, the 'mobile' and the 'laptop' have become near-universal machines. Both are devices for a multitude of trivial and advanced tasks in daily life. In this, personal media enhance social change in that they accelerate certain changes and compensate for roads not taken. This has always been the case with media technologies – what is new is that these forms of social regulation have reached the private realms of the individual in a much more fundamental way. Social interaction to a greater extent than before rests on the responsibility and the possibility of the individual, and is shaped by the time–space features of media.

Rather than focusing on distinct media with their distinct functions and purposes, a media-pragmatic approach addresses communicational selectivity and practicability in the constant social–technological interplay which establishes routine preferences and choices by skilled social actors. Personal media belong to the social lives of people who are much more than 'users'. The person and their personal media knit a complicated web of meaning, where face-to-face contact plays the most prominent role. Along with face-to-face interaction, personal media are entangled and continue to approach one another and our various social ties in our individual lifeworld. From a media-pragmatic perspective, this range of media alternatives and supplements are channels not only to a wider world, but also in the personal world of meaning-making, and of things present and absent, practical and impractical, convenient and inconvenient, impulsive and calculating. They are, as language itself, non-human, but still very human-made tools for constructing and maintaining reality.

DOI: 10.1057/9781137446466.0005

5
Personal Media Theory

Abstract: *This chapter takes the step from communication to media technologies. It briefly revisits Marshall McLuhan's ideas and some other proposals in the medium theoretical tradition in order to grasp the power of media technology and to what extent it can be 'tamed' or domesticated. I address both the discomforting aspects of dealing with virtual (simulated) realities and on the other hand, the convenience they offer to our sociality. It also charts the debate on the relationship between the online and offline in everyday life, and illustrates with the very recent development of augmented media/situated simulations.*

Rasmussen, Terje. *Personal Media and Everyday Life: A Networked Lifeworld.* Basingstoke: Palgrave Macmillan, 2014. DOI: 10.1057/9781137446466.0006.

Apparently, personal media require a higher degree of involvement from the users in interactive and immediate dialogical processes. The difference from the mass media is often conceived of as one of the different levels of commitment on the part of the user. As the *sender – medium – receiver* model becomes obsolete, personal media require more determination about the intentions of the process than the more habitual media reception. However, this intuitive 'observation' masks deeper and more significant social and cultural differences between action orientations and kinds of media. Personal media use should be approached hermeneutically, both from a typology of everyday genres and from a rough typology of technologies.

Nancy Baym (2010, 7–12) suggests that media as well as face-to-face communication can be compared along seven features: interactivity, temporal structure, social cues, storage, replicability, reach and mobility. The mobile is primarily an extension of hearing and voice capacities. The relations in space (and with answering services: time) are fundamentally altered since I may not even know where the person on the other end (the object) may be located geographically (or in time), and this gives me the reason to ask: 'Where are you?' The object may be reduced to a voice (even with unsatisfactory technical quality), that is both 'here' and 'there'; it operates in a partly irreal near-distance (Ihde 1990, 78) – in a self-generated *virtual context*. The mobile phone seemingly narrows the distance between the subjects, by 'magnifying' the other. A spatial, contextual change takes place in that the other comes closer to one's bodily context. The telephone *itself* 'withdraws' to clear way for the enlarged view of the other who becomes more 'present'. Through this kind of use, it is a tool for a moment of intersubjectivity. The phone is no longer an object, but refers to a human *co-subject* that must be reached in a process of meaning-in-constitution (Rasmussen 1996).

As such, the smartphones are taken-for-granted, 'naturalised' facts of daily life. This demonstrates the necessity to view media technology hermeneutically, to notice its contextual relationship to the subject, how it is constituted through its use. New means always imply modified uses and perceptions. Despite their transparency or even triviality, the mobile and other personal media transform perception by isolating, framing, amplifying and revealing (new aspects of) the object, and also through the relativisation of distance. They enable production of meaning dialogically. In the case of the telephone it is important that one hears (however

DOI: 10.1057/9781137446466.0006

modified) an *actual* voice in time and space. The pseudo-transparency of the mobile implies that *one hears through it, not at it.* The medium is 'close' to the subject, that is, it is a tool for the subject in his/her contact with the world. The technology is 'here', compared to the other out 'there'. Also, in a telephonic conversation, the other becomes a 'co-subject', because of the converging of utterances into one communicative process.

These facts about telephony stand in contrast to the relation between the human subject and mass media like television (TV, or the radio). The TV becomes an object of perception *in itself* while it also refers to something else in the world. What is immediately interpreted is the technology of sound and images – not the world through it. We watch 'TV news', rather than 'world news'. With the TV, the focus of the subject is directed much more towards the terminal than in the case of the telephone. Similarly, while the telephone is predominantly a channel for interaction, the TV and radio *themselves* constitute objects of mediation. Although the technology is located in the same locale as the viewer, its significance is 'out there', merged with the world of events and people, which it mediates into the locale. TV presents dynamic, audio-visual representations of the world rather than verbal interaction with co-subjects in a discursive mode. By contrast, if the telephone presents the object in a distorted way (supplying the object with a false voice, etc.), it is due to changes of the other agent herself or to error committed by the subject, and not caused by the interpretative use of the media technology itself. In the case of the telephone, one perceives the other subject *through* the technology, and thus gives the technology relatively few opportunities to present cultural versions of the world.

The point here is not that all media behave as either the mobile or as the TV. The web is the most important example of a medium that transcends the distinction by importing and integrating communicative and hermeneutic features. Various media entail different weights of the communicative mobile aspect and the hermeneutic TV aspect. Most media technologies mediate the subject–object relationship differently, and so locate themselves on one or the other 'side' of that distinction. From an interpretative position of the agent/user/subject, all media technologies can be placed on a continuum between the two extremes. The location of the web would depend on its interactive design and use of sub-media like email, mailing lists, chat groups and so on. It is a question of allowing the reader to write.

DOI: 10.1057/9781137446466.0006

McLuhan

Marshall McLuhan, as we know, was a media thinker of many fascinating, half-baked ideas. A perspective that ran through his work was present as a turning point between the old and the new; the media in a waning world of distance and a coming unifying world of inclusion and integrated meaning. However, below the grand ideas, do we find traces of a theory of the new personal media? The revival/return of an acoustic, tribal collectivist world as an effect of telecommunications, broadcasting and heretical ideas in science was touched upon in the concluding chapter of *Gutenberg Galaxy* (The Galaxy Reconfigured). In *Understanding Media*, this turn was analysed more specifically, if still in a very McLuhanesque and fragmentary manner: Let me briefly revisit McLuhan's seven rhetorically grounded (sets of) arguments or hypotheses about media and human change in *Understanding Media*, and then his four laws of media.

One: The medium is the message: '... the personal and social consequences of any medium – that is, of any extension of ourselves – result from the new scale that is introduced into our affairs by each extension of ourselves, or by any new technology' (1964, 23). This statement means several things. First, an argument about social/mental construction: that the formal or material composition of the medium will present the world differently, hence our perception of it. To understand this we should consider the media as extensions of our senses. Our consciousness will be shaped according to the formal features of our 'extended senses'. Second, there is no real content; all content consists of other media. The content of writing is speech, the content of print is the written word and the content of the telegraph is print. The use and content of media may be diverse, but it is also of little importance: '... the medium is the message because it is the medium that shapes and controls the scale and form of human association and action' (1964, 24). Third, our conventional emphasis on use and content tends to distract us from what is important about the media, and prevents us therefore from seeing the new age carried by the electric media: 'For the "content" of a medium is like the juicy piece of meat carried by the burglar to distract the watchdog of the mind' (1964, 32).

Two: Media can be classified into hot or cold/cool media (or possibly listed on a continuum between the two) The radio is hot, while

DOI: 10.1057/9781137446466.0006

telephone is cool. Movies are hot, while TV is cool. The alphabet is hot, while hieroglyphs are cool. Waltz is hot, while jazz is cool. The principle refers to the ability of the medium to involve the user. While hot media provide complete information, cool media provide little or incomplete information that must be supplied by the user. Hot media do not leave it to the user to fill in information. A related distinction is between high- and low-definition media: High-definition media fill the user with information, often towards one or few senses, while low-definition media provide incomplete information towards many senses. Low-definition, cool, less-specialised media tend to encourage participation.

Three: At later stages of the development of a medium, the effects tend to reverse from the original effects. This can be seen in the development of singular media, and in the media culture as a whole. TV and telephone begin their careers as individualistic media, but tend to turn users towards each other. With computer power, the typed word, which once was differentiated from the spoken word, now turns towards orality again. At the level of culture, tendencies of fragmentation switch towards growing social integration and 'retribalisation'. This point reappears in later publications, such as the posthumous Media Laws (see later).

Four: The media represent extensions of our senses, and precisely therefore also provide a sort of auto amputation since the media irritate some senses at the expense of others. As in the myth of Narcissus, we become numb or shocked by the composition of the media, which changes the equilibrium of the nervous system. The senses seek new equilibria depending on which senses become extended by the media. The introduction of TV is changing culture and personality everywhere, depending, however, on the existing sense ratio in each culture. For instance, in the relatively audio-tactile Europe, TV is intensifying the visual sense, and therefore making Europeans more similar to Americans.

Five: Media change must be understood as an evolution where media compete, influence and change one another: 'The interplay among media is only another name for this "civil war" that rages in our society and our psyches alike...The crossings or hybridizations of the media release great new force and energy as by fission and fusion' (1964, 57). On the one hand, a war is going on: 'When the

DOI: 10.1057/9781137446466.0006

press opened up the "human interest" keyboard after the telegraph had restructured the press medium, the newspaper killed the theatre, just as TV hit the movies and the night clubs very hard' (1964, 60–61). On the other hand, the new media constitute new relationships that change their ways: 'Radio changed the form of the news story as much as it altered the film image in the talkies. TV caused drastic changes in radio programming, and in the form of the *thing* or documentary novel' (1964, 61).

Also, this 'hybridisation', these moments of change, opens up possibilities to glimpse the structural properties of the media. Normally, media put us in a state of numbness that diverts our consciousness while the media themselves 'slam the gates of judgment and perception' (1964, 68). With the nationalism of the printing press as a recent backdrop, we may better observe the 'tribalism' of radio. Such media meetings cause ruptures, which again may enhance reflexivity: 'The hybrid of the meeting of two media is a moment of truth and revelation from which a new form is born. For the parallel between two media holds us on the frontiers between forms that snap us out of the Narcissus-narcosis. The moment of the meeting of media is a moment of freedom and release from the ordinary trance and numbness imposed by them on our senses' (1964, 63).

Six: Mechanisation is a translation of culture and personality. It used to mean translation into more explicit, specialised forms. With new technologies, we are translated into portions of information, which is much more integrated. Just as humans used to be servants of tools to survive, we are now dependent on new media. However, whereas old tools tended to be fragmentary and partial, new technologies tend to operate more inclusively and organically, and so encourage tribalisation/communality.

Seven: There is a possibility/promise that the new situation may rebalance the ratio of the senses, and so reset our rationality: 'A new stasis is in prospect' (1964, 68). All media change affects the entire system or ration among our senses. Just as our senses constitute an integrated system that changes when a sense damaged, new media that amplify or intensify the input from a sense change the entire consciousness and culture as system. There is however one immunity; art: '... the artist is indispensable in the shaping and analysis and understanding of the life of forms, and structures created by, electric technologies' (1964, 70). 'The artist is the man in many fields, scientific or humanistic,

DOI: 10.1057/9781137446466.0006

who grasps the implications of his actions and of new knowledge in his own time. He is the man of integral awareness' (1964, 71).

Having put forward these arguments more or less systematically with more examples than references to research, and with references to everything from Toynbee to Agatha Christie (and of course to James Joyce), McLuhan proceeds to discuss a wide variety of media in turn. In 25 chapters he addresses singular media from the spoken word to weapons, in relation to human senses and culture. His famous style is associative, metaphoric, filled with more assertions than arguments about media and culture, mixed with bewildering and thought-provoking examples and references.

In two overlapping works, collaborators of McLuhan have put together central ideas that McLuhan developed during the 1970s. The origins of the ideas are somewhat unclear; to what extent these ideas are developed by, or simply formulated by McLuhan's associates remains open. The tetrad is a set of four questions that can be asked with reference to all media and material things. The thinking is inspired by classic rhetoricians and others who have opposed dialectics from Aristotle to Hegel (and Marx), preferably Plato, Aquinas, Bacon, Vico and Joyce. Rather, what we have is an ongoing interaction between forms and background furthered by the media. New media bring new aspects of reality to the fore, while others move to the background. From the electromagnetic telegraph, media have tended to bring back auditive, tactile ways of perceiving the world, tipping our orientation towards the right hemisphere/part of the brain, and culturally, towards the acoustic space of involved, emotional social contact. The reader can either see the tetrad as a simple methodology for addressing media change, or (as McLuhan did himself) look at this as an alternative research paradigm concerning media change and the history of consciousness.

The four laws of media and their corresponding questions are as follows: What is extended or enhanced by a new medium (what becomes the new form)? What phenomena are made superfluous or redundant by it (what becomes 'ground')? What phenomena are retrieved or brought back (from ground to form)? What phenomena tend to end up reversed, with contrary effects (from form to new form)? Thus, all media and technologies will tend to enhance something, make something superfluous, recreate something and press something towards its opposite effect. The telephone enhances voice communication, makes the presence of the body superfluous in conversation, brings back the personal, mystical

DOI: 10.1057/9781137446466.0006

and 'telepathic' and flips over to phenomena like telephone conferences, which distance the voice from its intimacy. Electronic mail increases information transfer, makes interpretation of all mail impossible, creates new patterns of communication and taken to the extreme, will create chaos and loss of identity.

The first law/question is the McLuhan question proper, famous from *Understanding Media* (each medium will amplify or enhance distinct senses and aspects of reality at the expense of other aspects). The second law/question is basically the same question as the first, but with emphasis on what is pulled back by the new medium (e.g. print culture overshadowed by TV/film/telephone). The third law/question refers to the point that old innovations and things occasionally get a 'second life', however in a new modern, electric or digital form. What appear as clichés reappear in new shapes as new useful things. This is easy to see in fashion, typography, social rituals and all media, which all can be considered as modernisations of archaic forms of communication (consider also current concepts like 'vintage', 'classic', 'shabby-chic', 'retro', 'nostalgia', etc.). The horse-cab turns into a car-cab, mail into email, walls into Facebook walls and so on. The fourth law/question of reversal concerns unintended and unrecognised side effects: the car creates suburbs; the telephone creates skyscrapers; information overload creates illiteracy (?); cash money reverse into credit cards; bureaucracies create informal channels of communication and so on.

Not much used in media research, the laws of media are considered as determinism (no human influence), and as trivial compared to McLuhan's rather ambitious presentation ('The New Science'). Without doubt McLuhan's legacy lies in the first law, the central message of *Understanding Media*: TV and other cool, low-definition media take over for the hot, high-definition media, and are, through their acting as selective prostheses on our senses, re-enhancing a more interacting, informal, involving, Asian-like, global culture.

Modes of mediation

McLuhan's ideas carried further a media-centred perspective on media history that perhaps began with Harold Innis, and continued with Walter Ong, McLuhan and Neil Postman, Jean Baudrillard, Joshua Meyrowitz and others. A concept in this tradition subject to some debate is reme-

DOI: 10.1057/9781137446466.0006

diation. In the book *Remediation, Understanding New Media*, Jay David Bolter and Richard Grusin (1999) argue in line with McLuhan that new media convey old media as content. What they call *remediation* drives the media evolution. Remediation involves two dimensions or perspectives: *transparency* (or immediacy) and *hypermediacy*. On the one hand, transparency means that the medium withdraws itself from reality, to let the immediate, direct experience (visually and auditively) constitute the user's frame of experience. The medium hides itself, letting reality appear. On the other hand, hypermediacy derives from rejecting the idea of total mediation; it stems from graphic artificiality, tabloid layout, computer-game virtuosity and post-modern artificiality, in staging the object as unfinished, incomplete or as a model. The medium 'wants' to make the mediation explicit, and the user feels that she constantly needs to make decisions concerning what information to receive and how to receive it. The distinction between transparency and hypermediacy is a post-McLuhan way of connecting media to perception, but by exchanging hemispheres and brain functions with aesthetics. Bolter and Grusin's theory of remediation is frequently used to understand digital media.

In a similar fashion Fagerjord (2003) Fagerjord argued that each new medium carries with it particular rhetorical features or techniques, and that these are borrowed from previous (or existing) media. To understand the appearance of video, text and sound on the web, we need to address how the rhetorical forms or existing media forms are recombined in the new medium. Fagerjord's concept of 'rhetorical convergence' and Bolter and Grusin's terms 'hypermediacy' and 'transparency' all point towards different strategies in analysing conventions inherent in new media, deriving at least in part from the technology itself. For instance, in the multimodal process of remediation or rhetorical convergence on the web, both writing and images seem to receive new functions: As Bolter argues, the technology has the effect of making text graphic by representing its verbal structure graphically. On the web, the emphasis on design is radicalised. Java applets, icons and other clickable links give the writing a more visual, symbolic style.

Another useful distinction, to examine the various features of personal media, and web media in particular, is showing/telling as it is addressed in different research disciplines. Philosopher of art Susanne Langer's distinction between discursive and presentational symbols is a case in point (Langer 1942, Meyrowitz 1986, 95). Her theory of 'presentational symbolism' was influenced by Ernst Cassirer, and she argues that basically

DOI: 10.1057/9781137446466.0006

all forms of human expression could be seen as expressive, but that some were abstracted through signs/signals, and others abstracted through symbols. Symbols give expression to ideas that go beyond what can be expressed through language. A dimension of experience that cannot be accounted for discursively is dealt with non-discursively through a wide range of symbolic phenomena (myth, ritual, art, music, dreams). She also argued that iconic symbols of human emotions could be interpreted as rules similar to rules of language.

Discursive symbols are most importantly language, while presentational symbols are predominantly pictures and images. Discursive symbols are abstract and arbitrary in that they never mirror what they stand for. Also, they are discrete, which means that they carry a meaning independent of the immediate arrangement of the particular letter or word. Also, the grammatical structure of discursive symbols does not resemble the sequence and arrangements of how things actually take place. For instance, one cannot describe events simultaneously, even if they happen simultaneously in real life. She argued that many things did not fit the grammatical scheme of language, but were still of vital importance and of objective presence. Symbols take over when language must remain silent. They need to be conceived of through another symbolic schema (Langer 1942).

In contrast, presentational (non-discursive) symbols such as music, photo or video have a direct immediate resemblance with what they represent. The particular elements of a photo have no immediate meaning by themselves, only when arranged in a pattern that in some way represents objects in the world. While discursive symbols like sentences make sense only when they are organised according to an internal formal grammar and in clauses in discourse, thus independent of the actual world, presentational symbols make sense when they are arranged according to some actual 'real' external object. Consequently, the only way to make sense of a presentational symbol is to perceive it as a whole, in one act of seeing, as a gestalt. Unlike language, the photo represents itself and cannot be 'explained' by other presentational symbols. Neither can representational symbols easily be used to discuss abstract ideas and arguments without the support of discursive symbols.

Meyrowitz (1986) argues that Langer's terms may be used to specify the difference between electronic media and print media. Print media, on the one hand, exclude presentational meaning, leaving only the meaning of discursive symbols. Electronic media, on the other hand,

DOI: 10.1057/9781137446466.0006

convey a broader range of meaning in that they mediate presentational symbols along with discursive symbols. Radio conveys sounds along with verbal language, and TV adds images to verbal or printed language. Electronic media, because of their close affiliation with expressions and presentational symbols, '... tend to unite sender and receiver in an intimate web of personal experience and feeling' (Meyrowitz 1986, 96). It is worth noting that Langer did not think that presentational symbols express the emotions of the artist, but rather an 'idea' of emotions. They express illusions, virtual versions.

A similar conceptual dualism is presented by Watzlawic et al. (1967, see Meyrowitz 1986) who distinguish between 'digital' and 'analogue' symbols. Digital symbols are discrete symbols, as numbers and words, whereas analogue symbols are continuous and fluent. Watzlawic et al. argue that digital symbols convey manifest content about issues and things, whereas analogue symbols convey relational messages about feelings and other aspects that cannot easily be formalised in digital messages. Converted to the discussion of print and electronic media, Meyrowitz suggests that print media convey digital information, whereas electronic media convey both digital and analogue information.

These sets of concepts help to explain how contrasting media forms provide diverging possibilities and limitations for the media user in understanding the world. They convey different 'logics' and lead to quite different learning and interpretation processes. While presentational media encourage more personal, intimate and expressive responses, discursive media (text) appear more precise and 'scientific', but exclude feelings, impressions and personal undertones that unavoidably come to the surface in expressive information. Presentational media forms may present objects in an unambiguous and immediately understand-able way, but they cannot easily convey abstract ideas or conceptions in a precise fashion. To specify such symbols more accurately, one often must rely on language (as in subtexts of photos in newspapers). They are natural and direct in their representation of the world, but still imprecise and ambiguous. They cannot present statements that can be proven true or false. Rather, they appeal to taste and affections.

Personal media may mediate their meaning through both hot and cool media (McLuhan), immediacy and hypermediacy (Bolter and Grusin), through discursive, analogue texts and non-discursive (Langer) or digital/analogue (Watzlawick) symbols. Objects and places are non-discursive by nature; they are often products of art, craft and architecture. Ne

DOI: 10.1057/9781137446466.0006

simulation-oriented media, for example, let the viewer receive the story of the object 'told' by the object in its own presentational, symbolic ways of expression, only supplemented by written text (see later). Does this imply that the viewer comes closer to the object – with more open and direct access to the object since no information in another key comes 'between' the viewer and the object? Can presentational showing do the job of written text? In most contexts the answer is negative. The information that needs to be presented is normally too complex, exact and abstract.

Undermining representation?

This discomfort relates to what we may call fear of the Simulacrum: What is at stake is the status of representation itself, the process in which representation no longer respects the object, but replaces it. The original evaporates. The concept of *hyperreality* popular in the 1980s referred precisely to the experience of vertigo: That distinctions, spaces and time differences between objects and their representations tend to implode in the simulations and simultaneity of what Castells (1997) calls the 'spaces of flows' of the media society. The world of objects dissolves into an encompassing world of models. The model world takes over for both the real world and its representations. It creates a demand for itself due to its perfect illusions and for its reference to its environment as real. The notion of hyperreality (Baudrillard) refers to the perception of something that is neither real nor fake, a stage allegedly intimately bound up with a post-modern condition saturated with media, advertising and artificial environments.

Hyperreality refers to a stage of neither the real nor the representation, but something artificially simulated. A simulation is a representation that evolves to replace what is represented. It becomes 'reality'. Such simulations may be perfect but cannot avoid the emptiness of simulation and stimulation rather than true energy and 'aura'. It is a perfect imitation into, following Baudrillard's fourth stage, a simulacrum that transcends the real and its representations with no connection to reality. A double reference or a double loop is taking place: a simulation is pretending to be unreal. Unlike virtual reality, hyperreality creates a simulation within the real, which transcends the boundaries between the two into an apparently real reality. An example in Eco (1986), and Baudrillard (1983), is Disneyland. The land is perfect artificial realism (simulation), which

DOI: 10.1057/9781137446466.0006

makes it more stimulating and attractive than reality. This makes the term unreal meaningless and leaves the illusion that only what is outside the park is real. It is a world beyond reality and fiction, beyond true and false. Other examples would be TV reality shows and entertainment parks. The problem here is that representation was based on trust that the original existed as a guarantee, as a model. But if the original is a simulation, what is left is only circuits or systems where signs refer to signs. The fear is that the authority of originals is absorbed by the hyper-real, by the Simulacra.

Situated simulation

What has been called 'augmented reality' aims at presenting a totally mediated reality, since all traces of the physical world are mediated through the medium. Icons and text are laid over objects in the physical world, as an intermediary between the user and the physical world: 'Augmented reality is hypermediated, for it makes the user aware of computer graphics as medium, even if the goal is to keep the graphics and the external object in a close registration' (Bolter and Grusin 1999, 216). This is the dream of the smart-house and the virtual classroom. However, as a consequence of the fact that the dream of the virtual world is cancelled, other more hybrid versions are developed that account for the merge of 'the actual and the virtual' into a real world of a wide variety of mediated and unmediated experiences.

Let me illustrate this point. What is called a *situated simulation* differs from both augmented reality and virtual reality (Liestøl 2009, 2011). A situated simulation requires a broadband smartphone with substantial graphics capabilities and hardware sensors for positioning and various forms of movement. In a situated simulation there is approximate identity between the user's visual perception of the real physical environment and the user's visual perspective into a 3D graphics environment as it is represented on the screen. The relative congruity between the real and the virtual is obtained by letting the camera position and movement in the 3D environment be determined by the positioning and orientation hardware. As the user moves in real space, the perspective inside the virtual space changes accordingly.

In this modified augmented reality model, a distinction between the physical world and the virtuality of computer graphics does not exist in

DOI: 10.1057/9781137446466.0006

the medium, and so it does not aspire to capture all aspects of a physical/ virtual world into itself. Instead, in its mediation of reality, a situated simulation inserts a distinction between inside and outside the window frame of the mobile device. It acknowledges the ontological distinction between physical and virtual dimensions by letting the medium itself reproduce the distinction. The reality is 'augmented', not by capturing it entirely in the medium and then enhancing it by a graphical layer or filter. Rather, the world is here 'augmented' only by letting the medium mediate a virtual window, surrounded by a physical context. The unmediated reality actively takes part in the 'augmenting'.

This design avoids both the total rejecting of the physical world as in virtual reality (virtual transparency) and the excess of much ubiquitous computing, which celebrates an entirely mediated world (virtual hypermediacy). The medium does not attempt to take control over the world of perception – it is not its 'intention' to provide complete information about the world through (actual or virtual) mediation. It is a 'cool' medium. It respects the physical world as fundamental for situated experiences. Unlike the TV viewer or the user of VR, there is an enduring, material object (environment?) to observe. The medium obeys the fundamental interchange of perspective (ontologically and practically) between the physical and the virtual.

The medium mediates through a binary code we may call physical/ virtual. By being confronted with the object in a physical and a virtual version (coinciding with the unmediated vs. the mediated), and *de facto* observing the object through the two perspectives, the user will have to be habituated to perspectival shifts as procedure. By juxtaposing the two forms of mediation into a meaningful practice, the medium offers 'binary observation'. But it also enters what it mediates. By obeying the mobility, movements and the subjective perception of the user, the object can be perceived from various directions and angles, however in a virtual and trans-historical dimension.

Situated simulation is not intensively hypermediated as Bolter and Grusin (1999) argue, because it has no ambition to encapsulate reality. Rather, it mediates dualistically (physically and virtually), which is based on a balance between the transparent and the hypermediated. Bolter and Grusin argue in line with McLuhan that new media are representations of older media in another medium. They consider this to be a defining feature of new media. Such representation can however never be entirely complete or successful. The new medium always becomes dependent on the old, in a new shape, like the encyclopaedia is represented by and transformed by an online version – which nevertheless relies on the encyclopaedia concept of presenting definitions, descriptions and so on. Following Bolter and Grusin, remediation is what allows for new media, and at the same time what stops them from being complete or pure. Since the two dimensions of mediation in our case coincide with the inside and outside of the medium (but inside the experience), it is possible to describe them separately.

On the one hand, a *transparency perspective* lies in withdrawing its mediation from reality, letting immediate, direct experience (visually and auditively) constitute the outer frame of experience. On the other hand, a *hypermediacy perspective* derives from its pulling back from the idea of total mediation in its graphic artificiality, its computer-game realism, its emphasis on the object as unfinished, incomplete or as a model. The user experiences herself as someone who needs to make a series of decisions concerning additional (textual) information. Hypermediacy stems from text 'balloons' and audio, which provide comments (Liestøl, Rasmussen, and Stenarson 2011). Finally, it stems from the presence of holding the medium while standing/walking and its very visual distinction between physical and virtual reality.

When successful, situated simulation transforms this dualism into a duality, that is, into an augmented observation. This does not imply the complete elimination of the distinction between the physical and the virtual, but rather that the two perspectives refer to one another in a harmonic (user-friendly) way in an actual observation context. Whereas remediation is denied by the transparency of direct perception, remediation is celebrated as the main aesthetic experience by the hypermediacy of it. Although surrounded by transparency, the medium and its mobile graphics lead to the conclusion that situated simulation is a predominantly hypermediacy medium or genre (form of representation?)

As stated earlier our system departs from both virtual reality and most augmented reality systems in that it encourages the distinction between on-site reality and the computer-assisted simulation. For this reason our field trials focus on precisely users' experience of this distinction. This focus derives in part from an epistemological idea about the construction of meaning through differences. Our ambition is that the meaning of the experience does not derive from the simulation of the object, but from the distinction between simulation and the direct observation of the object. However, our intentions are worth little if the users

DOI: 10.1057/9781137446466.0006

feel seduced or alienated under the abstract spell of the simulacrum (Baudrillard 1994). Our field trials and user workshops test to what extent the difference-principle actually works as intended, to what extent the unmediated reality serves as a reality check vis-à-vis the simulation. However, equally important is to learn whether the simulation is sufficiently sophisticated as to enter into an interaction with the real. Much more than letting the simulation of the real undermine the *real* real, we search for a way closer to Lewis Carroll's map that had 'the scale of a mile to the mile': By letting the real and its simulation constitute a double description, the real is getting a chance to guide itself *about* itself, simplify itself and provide information about itself.

This follows the line of the well-known ideas of Gregory Bateson (1979). He draws on a wide variety of ideas from systems theory, biology, pragmatism and social anthropology. Bateson addresses how to identify homologies beyond scientific categories (similarities between a lobster and a crab). His famous phrase 'pattern which connects', refers to ongoing interaction between ways of perception, producing a rich vision and context for meaning. His relational view places conventional science in a context, as a rational way of perception, but it cannot avoid referring to its perception (itself) and can never prove scientifically. Information recursively feeds back to previous observations of an object made by the system, potentially correcting or calibrating classes of observations. Information emerges as a result of interaction between two sites or viewpoints, supplying depth and width to the observed, as in bilingual view, bilingualism, and being trained in two related disciplines and so on. The system may adapt its view of the object depending on its degree of flexibility and stability.

Oscillation between originals of the past and synthetic information of today may give experiential insight into an object that can no longer present itself in full because of time. Augmented reality provides double descriptions by adding a model world to the physical world – a twin world for adding perspective and comparative knowledge. The viewer now has two sources of information, which converge or fuse to form a single, however rich, object, an 'extra dimension of seeing' (Bateson 1979, 70). This however is only possible if true oscillation takes place. If successful this medium may help the user to establish an interchange between the real and the synthetic so as to create another dimension of information 'of depth', beyond what is actually presented. If they are accepted as exemplars of the same type, that they disclose different

DOI: 10.1057/9781137446466.0006

descriptions about the same type, they belong together and may be combined to enrich or enlarge one's view on the phenomenon. Double descriptions, if successful, create a *third* description, a synthetical inferential mode of information.

Convenient media

The network development, improved user interfaces along with further individualisation, brought media change to the current stage of digital personal media. The main forms of online activity tended to be chatting, through emailing, chat rooms, the mobile and SMS and messaging services. Increasingly, software allowed young people to create content for web-based sites. The change was quick, research of advanced personal media user groups ended up as relatively representative sample of young people in Western Europe. What were minor and peak phenomena at the beginning of the millennium is now mainstream. The affordability and improved user interfaces of multi-functional mobiles, laptops and mini-computers increased the competence and consciousness of both technologies and services on the Internet.

From the mid-1990s a series of media appeared on the Internet as channels of self-performance and self-presentation. The key to them all was management of personal information and social relationships. Now users present themselves in distinct and purposive ways that in accumulated form appear as social databases. Interaction of different sorts among users are stimulated and organised. There is little or no need for specialised computer or network competence. Along with creativity it is rather channelled towards writing, photography and video filming, which now can be made available in public spaces like Youtube.

As digital personal media proliferated, the avant-garde (that early research was based on) became a minority. Young people with digital media experience brought their everyday life into their new mediated forms. The new media were not simply alternative channels for interaction and expression; they were media for the very social life they lived among friends, family and associates. For instance, Lin Prøitz demonstrated how young people use mobile text messages and camera-phone images in their performance of sexuality and gender relations (Prøitz 2005, 2007; Lüders 2007). The same goes for the use of mobile telephony that exploded as a tool for coordination of daily life from the mid-1990s.

DOI: 10.1057/9781137446466.0006

Its use was predominantly associated with maintaining existing social relationships. We are thus not talking about separate realities and identities, but contexts of online and offline communication combined. As a general rule therefore, online activity should be seen as interdependent with offline life. Blogs, for instance, appear as individual, even intimate forms of self-expression through text, photos and videos, but are also a way of maintaining social interaction with others. Users create profiles dedicated to online and offline friends and visualise networks of relationships, all of which have real social effects.

The mediated and the unmediated are not different ontologies, nor are they different social spheres. Various media influence communication and interaction in different ways, but they all support the individual in his/her everyday interaction and meaning-production. On the one hand, unmediated face-to-face interaction is probably considered to have a prominent, ideal–typical status among most people. On the other hand, there is no reason to consider the unmediated as a pure or ideal form, isolated from the diversity of media use. Mediated interactions are as real as unmediated communication – it is influenced by both in terms of interacting persons, topics for communication and as a way for friendships and relationships to develop (Tanis and Postmes 2003, Brandtzæg and Stav 2004).

Viewing everyday life in a social network perspective helped to see this. Barry Wellman, Carolyn Haythornthwaite (1998) and others have pointed out that personal media support personal networks among friends and family. Seeing social relationships in a social networks perspective is an analytical key that neither exaggerates the role of distinct media, nor views social relationships as virtual communities. Personal media supports distributed patterns of communication in social networks, and people begin to use such media for social reasons. The more social relations one tends to have, the more diverse use of personal media is likely to be found (Haythornthwaite 2002). The stronger the tie, the more different media are in use in a particular relationship (Haythornthwaite and Wellman 1998). Thus, media choice is to a large extent influenced by the nature of the social relationship (Licoppe and Smoreda 2005). Or in a media-pragmatic perspective, relationships are formed by the social and material conditions at hand.

As we have reached well into the second decade of the new millennium, new technologies for perception, interaction, publication and archiving have been absorbed to a large extent and embedded in everyday

DOI: 10.1057/9781137446466.0006

practices. To most people in countries far beyond the developing part of the world, the mobile telephone has become simply the 'mobile', and the personal computer is the 'laptop'. Both are devices for a multitude of obvious and mundane tasks in daily life. They are themselves of less interest as their essential use is related to functions and services on the Internet and telecom networks.

At this stage, the media research agenda is society itself, as it takes different directions through its use of mundane media. Technologies and society are fundamentally intertwined and integrated. But at the same time we need to keep the two concepts analytically separate in order to see how technologies influence everyday life and vice versa. The view here is that personal media enhance particular forms of social change in modern society, and that they both accelerate certain changes and compensate for roads not taken. This has always been the case with media technologies – what is new is that these forms of social regulation have reached the private realms of the individual in a fundamental way. Social interaction is to a greater extent than before the responsibility and the possibility of the individual. The freedom and responsibility of mediated communication belong to most people in the world.

Rather than focusing on singular media with their distinct functions and purposes, a media-pragmatic approach addresses communicational selectivity and practicability. With this I refer to the constant social–technological interplay, which establishes routine preferences and choices as a pragmatic, mundane everyday practice by skilled social actors. Personal media belong to the social lives of people who are much more than 'users'. The person and their personal media knit a complicated web of meaning, where face-to-face contact plays the most prominent role. Face-to-face interaction, the fixed telephone, the mobile, the email, SMS and a wide range of other media are entangled and continue to approach one another and our various social ties in our individual life world. From a media-pragmatic perspective, this range of media alternatives and supplements are not only channels to a wider world, but also in the personal world of meaning-making, and of things present and absent, practical and impractical, convenient and inconvenient, impulsive and calculating. This is why much research on the use of personal media tends to report observations with limited generalisation value.

DOI: 10.1057/9781137446466.0006

6
Social Capital and Social Media

Abstract: *This chapter examines the accumulation and dislocation of social capital in everyday life and the intermediary role of personal media, particularly social media. The concept of social capital, it argues, helps to achieve a nuanced view on mediated social relationships in everyday life. Various approaches on social capital are addressed (Bourdieu, Coleman, Putnam, Granovetter and Burt). A balance between a conception of the 'undersocialised' and 'oversocialised' individual is examined through the concepts of skills, social capital and social network. It refers to research that illustrates the relevance of the concept in various situations and for distinct types of media.*

Keywords: communication theory; digital media; everyday life; lifeworld; media theory; personal media; social capital; sociology

Rasmussen, Terje. *Personal Media and Everyday Life: A Networked Lifeworld.* Basingstoke: Palgrave Macmillan, 2014. DOI: 10.1057/9781137446466.0007.

DOI: 10.1057/9781137446466.0007

People not only enjoy being in touch with others, they also gain from it, intentionally or unintentionally. Increasingly, social ties are a source for resources of various kinds and increasingly, these ties and connections are digitally mediated. This chapter aims at broadening the understanding of personal digital media in everyday relationships with the aid of the term 'social capital'. With Portes and many others, I generally understand the term 'social capital' as the ability to secure benefits through membership in networks and other social structures (Portes 1998, 8). I will provide a definition relevant for personal media later in the chapter. Suffice to say here that social capital refers to features that appear as resources for individuals and groups, and that are produced collectively. The term may also refer to the 'energy' that stems from more or less stable social relationships, and that enables the individual to reach certain more or less shared aims. It emerges from the structure of social relationships, but appears as individual characteristics.

In this chapter, I address the significance of personal media in everyday social networks by making use of the notion of social capital to make sense of people's media priorities, uses, and non-uses, as well as their views and norms regarding their mediated relationships. Skills and access related to the use of the broadening range of personal media play an increasing role in the shaping of young people's social life. With the growing diversity of personal media, the how will be more important than the how much: qualitative priorities concerning media and people will be at the centre of our interest, rather than time spent. If specified well, 'social capital' is a suitable concept in examining this, because it addresses the relationship between power, skills and social networks in everyday life.

Generally, the use of digital personal media in social networks (from mobile telephony to social media) tends to support social capital of groups, because such media allow for keeping social relationships in spite of individualisation and a hectic and mobile everyday life. (In certain cases, they actually reduce social capital as in the case of computer-game addicts, who pull out of important social networks.) In general, however, it is difficult to identify causal relationships between technology and social capital because social networks are virtual-actual: face-to-face contact and media contact interweave contexts, intersect and maintain social ties together.

A broadening repertoire of personal media provides the individual with a varied set of alternatives for interaction. Personal media have simply become decisive in achieving the values and objectives that we

DOI: 10.1057/9781137446466.0007

appreciate in life. The mobile and Internet media, with their features so very suitable for acting out one's (culturally informed) will and decisions, have become artefacts of welfare. Social capital has increased as a consequence of Internet-based social networks. What Lin (2001) calls 'cyber-networks' are constructed through email, chat rooms, blogs and so on, which have experienced astonishing growth. To the question of the implications of cyber-networks' growth for the study of social networks and social capital, Lin's short answer is: incredible – we are witnessing a revolutionary rise of social capital: 'In fact, we are witnessing a new era in which social capital will soon supersede personal capital [in face-to-face interaction] in significance and effect' (Lin 2001, 214). Quite clearly 'cyber-networks' mediate social capital in various ways. A simple exchange through email with another person may create weak ties. Network sites specialise in establishing latent contacts, which then can be actualised when appropriate. From social interaction on the Internet, numerous campaigns and movements have emerged, as have entrepreneurial start-ups in the new economy. And not least within the sciences new constellations are formed that produce new published knowledge. Through the Internet, access to people and information is possible with less effort. More informal relationships are possible, which thereby enable cooperation. Personal media and mediated social networks help everyday interaction to go on in spite of a hectic life of obligations and demands, and at the same time reproduce the stability of the network (Sproull and Kiesler 1991). It is the network aspect which is particularly important here. Danah Boyd (2011, 39–43) understands social media like Facebook as a genre of network publics. Network publics are publics that are constituted by network media in that they are constructed space and an imagined collective that emerge from the intersection of people, technology and practice. This hybrid term is meant to refer several functions that cover both networked and audience aspects. Individuals are allowed to construct profiles, produce a list of connections, and view other viewer's profiles and their list of connections. All these can be contacted interpersonally of in a (semi)public fashion. The main features are thus profiles, friends and public connecting features along with stream-based updates. Such features gradually form the basis of important aspects of everyday sociability. They form the infrastructure of social habitual contact. Of importance here is how they are suited for obtaining particular resources from particular others in order to harvest particular benefits in particular contexts.

DOI: 10.1057/9781137446466.0007

Elements of social capital

In addressing social capital, observers generally tend to distinguish between structural aspects, individual aspects and the benefits in question. Some emphasise norms and trust, others focus on the individual as agent. For instance, Lin (2001) argues that social capital has three segments: (1) individual position in networks (causes), (2) access and use of resources (processes, skills) and (3) the effects or outcomes. Naahpiet and Ghosal (1998) suggest three dimensions (structural, cognitive and relational) of social capital: (1) patterns of ties in networks, (2) personal relationships of individuals who influence their actions, motives, aspiration and so on and (3) resources that provide shared representations and norms, and systems of meaning in the network. Similarly, Huysman and Wulf (2004) distinguishes between structural opportunity (who shares knowledge and how), cognitive ability (understanding) and relation-based motivation (trust, reciprocity, respect). Portes (1998, 8) distinguishes between the actors possessing or claiming social capital, the sources of social capital, which is also a source of social control, support and trust (the group or network), and the resources or benefits themselves, usually information in some form. Licoppe and Smoreda (2006) argue that we need to keep two forms of complexity separate: the contents, genres, ties and formats of interaction, and the technical means for which such interactions can be used.

As noted by Ellison et al. (2007), social capital is not something an individual (or group) possess (as is the case with financial, cultural, human capital) but rather something that emerges from interaction and communication in some form of social organisation. Therefore the term is social capital. That is why the term can refer to both individual and collective processes. Social capital stems from the dynamics of social organisation, for instance, a social network, and is identified at the individual level. Social capital is an abstract concept in that it can be observed as something more specific like information, influence, goods, high status and so on. Also, as it derives from membership in a greater social entity, its structural aspects are of some importance, particularly in our case its technological mediation. With Huysmann and Wulf (2004), I see the technological alternatives applied by the social network as interdependently embedded in the network. The choice of medium is in effect often a question of choice of interaction partner and context, and

DOI: 10.1057/9781137446466.0007

vice versa. Social interaction is pragmatic, flexible and malleable, and carries with it rationalities about suitable media for various occasions.

I define social capital as the outcome of resources shared through skilled membership and position in a social network. Thus, I distinguish between three elements of social capital:

1   The (structural) social organisation (the social network structure) including personal media, which temporally, spatially and otherwise influences capital distribution and accumulation. This structural component relates to technologies, demography, transport and much more. Regarding personal media, questions regarding their handling of meaning as texts, sounds or images, their dialogical character, or their interconnection with other media and so on are of interest. Media studies and related disciplines would emphasise this element.

2   The skills and positions of singular nodes (individuals or groups) of the social network. The study of human nodes and their position in networks refers to the nature of embodied personalised or organisational power in particular positions vis-à-vis others, their size and centrality in the network, and the strategies applied in order to accumulate resources.

3   The actual resources that the nodes may receive as a result of membership in the network and that are used and converted into social capital, which intentionally or unintentionally leads to benefits, which are then reinserted into the network and so contribute to its reproduction. We may draw on Lin (2001) who argues that network media enable the accumulation of social capital in the form of several kinds of resources:

   (a) Information of many forms, for instance, about job vacancies as was demonstrated by Granovetter (1973) in face-to-face encounters or in dedicated media networks such as LinkedIn.

   (b) Influence, such as in lobbying. Political party networks are another example, where party membership leads to party network access and potential influence.

   (c) Credibility and reputation: In sociology, this was probably first commented upon by Weber, who in his writings on religion in America refers to a man who wanted to open a bank found it necessary to join the Adventist community to achieve high status and credibility in the bank. Still membership, whether formal or informal, in certain organisations in politics, charity, sports and so on, leads to a rise in respectability.

(d) Self-recognition: As we have seen, Michel Foucault (1994, 223) writes about Technologies of the Self (such as the notebook and the letter), which assist the individual in 'self-caring', that is, in reaching a morally accepted level of self-esteem and self-identity. Personal blogs in particular would be a good case in point in that they constitute the Self as an object for reflection in and through.

(e) Reciprocity: To these, I would like to add a fifth kind of resource, which not only underlines the fundamentally human need for appreciation and trust, but also focuses on the individual contribution to qualitative dimensions of the reproduction of the network. The observed needs to respond to others, appearing from other's requests, regards and so on. This corresponds to Malinowski's observations that receiving gifts creates debt, an obligation to give back. All personal media dedicated to dialogue in some sense would encourage – if only passively, reciprocity, due to the low threshold for response. In social media, technology enters in assisting reciprocity more actively, in that the system provides hints, reminders and notices, which trigger new responses from the users.

Capital in personal media

The term 'social capital' designates several basic sociological insights, but at the same time it needs to be specified according to its empirical use and context. Several studies a decade back questioned whether use of the Internet can enhance community involvement (Kraut et al. 1998, see also Shah et al. 2001, 141). Data from United States and the Netherlands suggest that users of the Internet not only seem to have a more active life than others, but they also travel more and work more. Activity leads to more activity (Robinson and Haan in Kraut et al. 2006). In the late 1990s, Kraut et al. (2006) argued that Internet, like TV, implies physical inactivity and limited social face-to-face social interaction. They found that Internet use leads to declining sociality and increasing loneliness and depression. Most such studies, however, did not focus on particular forms of use. The Internet was viewed as

DOI: 10.1057/9781137446466.0007

one single medium, hiding distinct forms of use with widely different social effects. As we know today, 'internet use' is not a much better term than 'technology use' as it presents a far too crude and undifferentiated view of social activities.

A growing focus on social capital has helped to understand under what conditions specific kinds of Internet use are related to social trust and civic activity (Shah, Kwak and Holbert 2001, 142). Social capital often refers to how various forms of sociality help to generate cooperation and social change, and naturally the Internet in its distinct media forms enters as a vital factor, in that it can connect people and help them to acquire information and discover opportunities in conjunction with others. People with stronger social networks tend to be healthier and happier than others (Kraut 2006, 7). Time spent on the Internet does not reduce offline sociability (Gershuny 2003). Data from Blacksburg, Virginia, indicate that people who belong to more than one community group are better educated, more informed and more extroverted. They have higher levels of participation, trust and community attachment than others (Kraut et al. in Kraut 2006, 180). The notion of 'bridge' between social networks is crucial. In the hands of human bridges, they argue, the Internet is 'a tool for maintaining social contact and relations and increasing face-to-face interaction, all of which help to build both bonding and bridging types of social capital in communities'. And conversely, households without Internet access are generally those who are already disadvantaged in other ways as well (Anderson et al. 2007, 40).

However, the ability to use and gain from Internet use is socially differentiated according to a number of external factors such as motivation and knowledge among parents and friends. Research also indicates that social ties created online are more fragile. People spend less time with, and are less committed to, online friends than to offline friends (Cummings, Lee and Kraut, in Kraut 2006, 266). Networks of relatives are more stable than networks of friends. Friends may shift status from central to peripheral because the nature of relationships is only decided by the involved. Family and friends offer the comfort of the familiar (Blokland 2003, 77) at least as much as confidentiality. Friendships may exist without involving one biography in detail for the other. They may exist in the present, for a present. Some friendships exist best as a two-person relationship while others thrive best within a group. Naturally such complex social–psychological features would decide the use of personal media for social contact. While social media and email may

DOI: 10.1057/9781137446466.0007

be used to involve others, the mobile conversation and SMS are better suited for twosome relationships.

Shah, Kwak and Holbert (2001) addressed Internet use according to four types of use patterns and different facets of social capital, and found that use of the Internet for social recreation (e.g. participation in chat rooms and game playing) was consistently and negatively related to engagement in civic activities, trust in other people and life contentment. In contrast, usage of the Internet for information exchange had positive impact on the same variables (Shah, Kwak and Holbert 2001, 149). Their decade-old results suggest that the connection between Internet use and civic activities, interpersonal trust and life contentment was highly contextual.

The actual capital-enhancing role of the Internet use is very difficult to identify, precisely because it is so embedded in social life. It also seems vital to distinguish between forms of use along and with degrees of use. In neither case is the Internet a causal agent, but a medium in all senses of the word – an intermediary variable between social networks and the individual, which may encourage or discourage forms of use. For example, trust refers to an essential worldview developed during a harmonious childhood and cannot be changed by the Internet. As Uslaner (2004) reminds us, trust refers to an attitude towards strangers established long before the first Internet experience. But Internet availability may lead to the exercising of trust. The net makes things easier for those with many friends than for those with few. As Uslaner notes: 'The World Wide Web is very much like the world. It makes things better in some ways and worse in others. But it is not transforming. If you want to make a revolution, you have to go offline' (Uslaner 2004, 239). After the Arab spring, this statement probably still holds.

The nature of the medium may invite specific forms of social interaction. Sproull and Kiesler's 'Reduced social cues model' (1985) were intended to demonstrate that reduced social cues produce a less gratifying communication experience, resulting in a de-individualising effect, and inducing behaviour that is more self-centred and less socially regulated. Examples would be racist statements and 'flaming' in chat rooms. Thus, the use of various media can in some instances be taken as a statement of closeness (Licoppe and Smoreda 2006, 298). Media activating the human voice without delay would signal intimacy, whereas email indicates a somewhat larger distance or thinner tie. Late replies are also indication of a relationship less close than a relationship where SMS or

DOI: 10.1057/9781137446466.0007

emails are exchanged with greater frequency. As Licoppe and Smoreda (2006, 298) noted, 'the differential use of particular means of communication thus lays down a space of relational practices in which ties of similar closeness are treated in a similar way, and in which this degree of closeness is publicly expressed and negotiated'. If an announcement is addressed to one or more recipients, whether it arrives via someone else and whether it is simple/short or complicated/long are other variables (Licoppe and Smoreda 2006, 299).

Of course, time is an inherent dimension of sociability, and the choice of technical media is a way to use time to organise sociability. And as more than one person is involved in this organisational activity, the process takes the form of a meta-process of negotiating time. For example, friends may have different expectations with regard to frequency of chatting and choice of media, which influence the dynamic structure of the dyadic tie or the greater network. And yet dyads or social networks involving more people will tend to order themselves in rhythms and workable social distances. Although social relationships may vary in closeness and frequency (friends may become distant contacts and vice versa), they seem to be relatively robust and able to reshape themselves. Most friendships operate on a second-order level involving at least some vague reflection on the relationship, which sorts out how the relationship should be maintained. Such more or less reflexive preferences would involve activities, interaction under what circumstances and contexts and with what media. The mobile, for instance, serves as a channel of trust in spite of mobility. As Deborah Chambers (2006, 151) argues, the mobile is the ideal technology for the exploration of intimacy, in the form of loyalty and privacy as well as solidarity. The mobile helps to enhance strong tie networks, based upon individualisation. It is network-oriented rather than community-oriented (Licoppe and Smoreda 2006).

Investing in the mobile as 'Link-up'

Among teenagers, face-to-face contact tends to be so frequent that mediated contact functions as extending the presence of the other. Among adults and especially after many years of friendship, a phone call or an email serves to keep the relationship warm. In both cases, physical and mediated togetherness are interwoven. The mobile is of

DOI: 10.1057/9781137446466.0007

particular importance in this respect. In all parts of the world, the mobile is an embedded tool for coping. Particularly among low-income cultures, connectivity is a condition for surviving. In many poor cultures, the very fact of being in active and negotiating relationships with others is a major source for income. In poor countries (and in contrast to the Scandinavian countries), social capital is more important than cultural capital.

In their analysis of the use of mobile in Jamaica, Horst and Miller (2006) demonstrate how what they call Link-up through the mobile has become an essential part of everyday life. The almost global spread of the mobile has in many places led to personal networks (ego-centred networks), which tend to become an essential form of sociality, at times at the expense of more communal forms of sociality. The mobile has become the main technological tool for personal networks, which contribute significantly to changes in sociality. The phone is used to actively be involved in the lives and troubles of friends and relatives, and to exchange favours and services, which make life easier. The mobile phone list not only contains the names of family, friends and acquaintances, but also those distant contacts who may help out with possible favours. The mobile is heavily integrated in the informal and black economy, which is essential for survival in many low-income cultures.

Horst and Miller also show that the mobile is used to keep social relationships separate and to influence flows of information. It is a tool for empathy, trust and care, but also for (within households) deceiving and administrating sexual relationships. Link-up involves, according to Horst and Miller (2006, 96), a high number of very short calls to keep in touch, and to accumulate a high number of contacts in the SIM card. As an essential form of social interaction in mobile, low-income societies like in Jamaica, the mobile enhances social networks that do not always have much meaningful content, but are essential all the same for coping. Link-up is a social pattern of interaction and demonstrates how the mobile has become appropriated into all aspects of everyday life of ordinary Jamaicans. The essence of the social role of the mobile lies not in the conversations as such, but in the ongoing overlapping complexity of actual and potential connections and encounters. Horst and Miller (2006, 101) call Link-up a genre which may involve a host of emotions (suspicion, love), but where each call is most essentially a series of confirmations of liaisons.

DOI: 10.1057/9781137446466.0007

Skills

As the Link-up case illustrates, social skills produced in and through the individual lifeworld play an important differentiating role. Skills concern how and to what extent one makes use of resources to reach certain ends and to perform in the reproduction of social networks. Social capital is structured through membership in a social organisation, and through the individual skills. Agre (2004, 202) argues that it is important to separate out social skills that to a variable degree enable the actor to take advantage of his or her access to the resources of the group. A notion of social skills, social competence, digital literacy, agency or simply *habitus,* refers to the actual operation of social capital as a social mechanism influencing inequalities and exclusions in status and information. Social skills transform resources of networks into social capital for the individual or group. Social skills and social capital are thus interdependent properties of the social network. It is the interplay between individual skills and network that builds social capital, and that subsequently may benefit both skills and networks.

The discussion has mostly addressed what mechanisms of social structure tend to give access to social capital, such as the nature of trust mechanisms, and the difference between open and closed networks. The term tends to ignore aspects that may explain the variation in the ways actors make use of their opportunities, such as cultural capital, social skills and moral norms. I follow Agre (2004, 202) and Coleman in that social skills are a form of capital of their own, usually coined human capital. Social capital is a resource for those who already possess the skills to gain access to social networks, and make use of the resources obtained from those. Conversely, access and central positions in social networks are a result of skills that are in part learned through contacts in social networks such as the family. In this way, the dynamic of social capital and social skills is a self-enforcing resource.

However, the obtaining of skills is also a resource that derives from wider social networks such as political communities or colleagues. This explains why people of excellence in terms of knowledge and competence are sometimes left to see others without the same qualities make careers.

However, skills also refer to the sensible operation of the actual personal medium like the mobile or the web. There are considerable differences in what functions and options we actually apply in our devices, and quite

DOI: 10.1057/9781137446466.0007

a lot of people refrain from using the web altogether. By social skills, we also think of the abilities to handle the available repertoire of personal media in socially advantageous ways on the part of the user and the network as a whole. This involves selecting specific media for distinct purposes and communication partners, and steering away from the social and ethical pitfalls that follow in the footsteps of all media. Social capital involves acting responsibly vis-à-vis oneself and the network. This means, for example, to reply promptly on SMS, to act civilly, to refrain from uploading intimate information and so on. Increasingly, the failure to handle our new wide-reaching personal media with care may have serious implications for the social life of the individual.

The quality and usability of a website for an organisation or association influences the nature of trust in the network, by sharing information, obligations, trust, norms, arguments, ideas and so on. This is often seen as elements of *literacy*. Particularly personal media literacy is of relevance. Personal media literacy refers to 'the interpretative and writing skills necessary to communicate effectively via online media' (Warschauer 2004, 117). This includes the internalisation of a range of informal and unwritten rules, from the netiquette of civil communication to the pragmatics of effective argumentation according to the topics discussed, the nature of the media and the participants. It includes how to write emails in an effective and civil way, and how to make a personal homepage personal but not embarrassingly private. In the case of homepages, blogs and digital storytelling, skills refer to the ability to compose appropriate narratives. But personal media literacy also includes avoiding flaws (such as using the 'reply to all' button when replying personally to a widely distributed email.) It refers to how to establish and continue online relationships for the benefit of oneself and others. Such mastering is normally learnt through various social networks and settings, from school, family and friends along with experimentation (trial and error).

Capital dynamics in social media

Regarding 'social' media like Facebook, LinkedIn, Google+ and so on, a number of social issues are relevant for research on social capital. These include issues such as *general media use:* concern about how people are incorporating social media in their everyday life, in relation to their activities and their use of other media, changes between age groups and life phases and so on, as a foundation to understand changes of everyday life. Also *identity* themes concern how social media are used as a looking glass, as a façade, as impression management and so on, and as technology of the Self for the care of oneself. It concerns the ways the individual handles pride and shame (along with rejection, insecurity acceptance, etc.) as fundamentally relational emotions.

In order to understand how social media produce social capital, I would first like to indicate how social media create social capital structurally, and then address how social media appear as a social system or circuit of activities on both individual and structural levels, making mediated social networks into something different than the sum of its members. Social media networks generate social capital through a number of structural features, often conceptualised as the 'architecture' or 'topography' of networks. From the insight of network theory, I will mention four such structural features drawn from network analysis (Rasmussen 2008):

> *Clusters:* By mediating clusters of strong ties, and with high density of connections along with unifying norms and trust, social media mediate small community-like groups of sociality and loyalty. The effect of such bonding groups is a high degree of homogeneity. Such groups provide direct links to others, who each have links out of the group. However, such groups are relatively introverted, with relatively few connections out of the group compared to the connections inside the group. In other words, the connections are quite far from random-like. The redundancy is high inside the group, but low out of the group. The kind of social capital from such groups (best described by Coleman) is also called 'thick' social capital. Such clusters inhibit small, world connections for the individual of the group. But when such clusters appear as *nodes in larger networks,* they serve a 'small world effect' from the local level.

> *Short-cuts/redundancy:* In networks of relatively random-like connections between the nodes, enabled by a large share of weak ties, bridges will also develop between networks. In large, random-like networks, there are fewer clusters and more random-like connections, and therefore a shorter path distance than in a world of clusters. For example, a large share of my connections also know someone I do not know, and some of them serve as bridges to other networks. The network is more individualistic, redundant and heterogeneous. It more easily connects different groups.

> *Supernodes:* Online and offline networks tend to generate a certain bias, favouring the nodes that enter the network early or possess other advantages. Thomas Merton called this the Matthew effect. The long tail thesis actually shows the same thing, although the argument is different. The web-topology

develops supernodes, relatively few extraordinarily popular sites, and a very large number of smaller sites. Web use creates 'mountain-peaks' in the horizon, which everyone sees regardless of where they are positioned, and which dramatically reduce the number of steps between randomly selected nodes. Power-law distributions emerge, due to knowledge, conformity and time. Herbert Simon pointed towards time and attraction as the two central mechanisms that tend to create bias. A self-enforcing differentiation takes place, which is strategically developed, as in Google's search method, and in rating methods (Slashdot, Digg). Popularity leads to more popularity.

Cascades: Increasing exposure of practices of others with lower threshold than oneself leads for socio-psychological reasons of conformity to lower thresholds for joining campaigns, petitions, spreading 'memes' (jokes, images) and so on. Snow-ball effects emerge and may appear in virus-like phenomena due to the speed and reach of information. This is particularly true for information, which seems important or provocative. As accumulated side-effects, certain forms of information are imitated and duplicated locally, and then escalate widely.

Successful social media allow for all these features in a productive combination. They help users to navigate locally and long-distance, to identify changes and pick up news and groups, and to establish new connections. They potentially magnify effects of individual actions and so instigate social change. But they all remain dependent on individual activity on the mobile or the laptop keyboard.

To see individual action and structural features together, and also with both social and technological features, we may see social media as a complex socio-technical system (of systems). We may distinguish between an individual level of social activity (publish profiles, click on likes, poke others, read others' profiles and other people's feeds about what they do or think now, post comments, hyperlinks, photos and so on on the group walls, accumulate friends) and technological operations (suggesting friends, groups, who to poke, birthday reminders, helping to search for friends, keeping score on requests, friends, updating profiles etc.), and a structural level of social (social control, trust, moral, critical mass) and technological phenomena (control, oversight, stability). Various forms of techno-social activity circulate between the two levels and between the social and the technological into a mediated social network (consisting of thousands of interconnected networks) of stable change.

From this, it is evident that social media like Facebook are not only spaces or sites for communication, but also in fact 'avatars', which interact with the users. The success lies in the combination of the similarity with the social world (communication) and the difference from it (effortless and compressed). The result is a world of social density worthwhile spending time in. Network media minimise/shrink/compress the social world by way of technological code. They make the social world of the individual into a convenient entity, almost a material unit that can be observed and handled in different ways. They combine the simplicity and oversight of the map with the social reality of the world. In integrating the model and the reality, social media visualise social patterns. The net result is more sociability with less effort (lower costs and thresholds).

Clearly the notion of unintended consequences plays a role here. Social media like Facebook can be considered as a 'social fact' in a durkheimian sense, as a durable, collective social phenomenon external to and influencing the individual, and still reproduced through individual practices. Social facts (material and immaterial) are sets of values and norms in a society, social community or social integration in various spheres of society or in society in general. These social facts can be identified through sociological method and vary among cultures and over time. The main point here is that the social fact is external to the individual, in the sense that it is an unintended (and often unrecognised) effect of everyday practices, which are to some degree defined by custom, tradition or norms. A similar sociological argument is Giddens' structuration argument that structures (I think to a certain extent comparable to Durkheim's social fact) are a medium and outcome of practices. However, as is noted by many sociologists, Durkheim overstated cultural authority over individual practices. On the other hand, Giddens tended to understate both collective and material powers. However, with different emphases, both enable us to think about social structures like Web 2.0 phenomena as a collective and unintended construction made by innumerable individual acts. The more social network sites depend on user-generated initiatives (as in the step from Web 1.0 to Web 2.0), the more essential this argument is to understand their stability and transformation.

In commercial Web 2.0 sites like Facebook, user activity serves individual and collective needs and interests, and provides value to adverts and the exchange value of goods they advertise by stimulating the demand for them. It seems that with Web 2.0 we have reached a new level in the gradual convergence of the roles as citizen and consumer. With the historical coexistence of individualisation and liberalisation, consumption is regarded as the crux of autonomy, as the epitome of individual freedom. Only the economical and political autonomous

DOI: 10.1057/9781137446466.0007

subjects can freely exercise choice among products and services. Also, unrestricted consumption is seen as self-expression and as an identity-building practice. Therefore, advertising may coexist with meaningful communication and interaction. Or conversely, in Web 2.0, highly efficient and flexible systems for private and public interaction are made possible by the commoditising and monitoring of the same interaction.

Web 2.0 services like Facebook make use of innovative technologies (Javascript, XML, Flash) that allow for a wide variety of social interaction. The concept of Web 2.0 refers to a combination of innovative technologies and social networks. As Cormode and Krishnamurti (2008, 1) argue, 'Web 2.0 is both a platform on which innovative technologies have been built and a space where users are treated as first class objects'. The main difference, they argue, is that in Web 2.0 the content creators consist of the majority of the users of the site. Design and features are created to make every user a creator of content, ranging from 'like' to uploading software. From 2007, other applications that can be added to user accounts in Facebook were supported through the opening of APIs. This opened up the possibility of widespread use. Cormode and Krishnamurti (2008, 9) suggest the following classes of practices on Web 2.0 sites, from the most simple and quick, to the more skilled: Clicks and connections (simple one-click actions, such as a rating, accepting a friend request, voting), Comments (adding brief responses and comments), Casual communication (sending messages), Communities (joining and interacting in larger groups) and Content creation (uploading original content, such as photos, movies).

Web 2.0 provides a simple user interface for a wide range of forms of interaction and expression enabled by recent advancements in web technology. Web 2.0 also refers to a particular innovation model, as it signifies an architecture that gives user-generated content first priority. 'This burgeoning phenomenon suggests that users are gratified in significant ways by the ability to play an active role in generating content, rather than only passively consuming that which is created for them by others' (Harrison 2009, 157). It seems logical to see Web 2.0 as a radicalisation of the 'participatory culture' (Jenkins 2006) of collaborative consumer communities enabled by the new media. As Harrison and Barthel (2009, 174) conclude: '... new media technologies now enable vastly more users to experiment with a wider and seemingly more varied range of collaborative creative activities'. The various features in Web 2.0 media constantly create incidents, changes, events

DOI: 10.1057/9781137446466.0007

and other novelties, which serve as new reasons for entering and interacting, which reproduce patterns of social interaction. From innumerable singular and trivial practices (notices, pokes, likes, greetings and so
on), the system generates large-scale patterns of movement. Individual
practices are transformed into collective and ordered subsystems of
walls and groups.

An aspect of Facebook, compared to other social media like Wikipedia and Slashdot, is its extremely decentred user structure combined
with a strong command centre for all technological and economic issues.
There is no central idea or ideology, no human moderation, no hierarchy
between ranks and no differentiation of status. This structure is implemented to reduce tension between individual freedom of expression and
collective interest in relevance and order. In Facebook, control features
are handled in decentred and technological ways.

How do technological and social aspects, software and meaningware
of Web 2.0 interact in order to generate so much activity? How is the free
rider problem solved? Web 2.0 partially avoids the Free Rider problem
by allowing for large number of free riders as long as the number of
'paying riders' is sufficiently high. Millions of free riders in, for example,
Slashdot, GNU/Linux and Wikipedia can be tolerated and do not erode
the system as long as the number of active participants counts in the
thousands. In fact, the free riders are considered as sensible users or
supporters and do not bother the active participants. This suggests that
Facebook either has solved the Free Rider problem or that the Free Rider
problem is not relevant for Facebook. Hetcher (2009, 996) prefers to call
Facebook a 'spontaneous order' and 'a spontaneous informal ordering':
It does not aim for one single outcome and is not a common project, but
still it is highly social.

The collective action is not a collective action mystery, because we
are not talking about altruism, but sociality (Hetcher ??995). Facebook
does not rely on peer production as in Wikipedia, since people tend to
approach places of sociality like the city market or village square. It is
simply experienced as in one's own interest to participate. The system
operates without being dependent on altruism or idealism, thus providing a more robust structure than for example Wikipedia. The normative
expectations towards the system are relatively low and thus also the level
of disappointments. For example, in spite of several well-known Facebook attempts to exploit privacy information commercially, people seem
not to defect – yet.

DOI: 10.1057/9781137446466.0007

As both a typical and exceptional Web 2.0 site, the interface technology of Facebook puts the user first in terms of interaction and creation (profiles, including comments from other users, simple connections to 'friends', simple ways to post various material and to rate other's posts, simple ways of creating groups, walls, public APIs and embedding of rich content (Flash videos), and RSS feeds and email (Cormode and Krishnamurti 2008, 3). Contrary to Web 1.0 sites (like Craig's list, Amazon, Slashdot etc.), Web 2.0 sites are entirely dependent on these features, and share many of the features of offline social networks. This also means that each individual user appears as the centre of the site universe, as is the case in the offline dimension of our lifeworld.

Jarrett lists the following interactive features of Web 2.0: *Flexible time* (in contrast to broadcast 'push' time), *Creative capacity* (tagging, uploading personal photos and music, creating profiles etc.), *Body–object articulation* (creative self-expression never fully controlled or shaped by the actual site), *Intensive use* (no exhaustive use of the technology – features like tagging and folksonomies help the user to make use of tiny, selected content of the site), *Concealment of expertise* (strategic denial of authority by the corporate owner to signal full authority and freedom of the user) (Jarrett 2008, 5).

A possible negative outcome of this is that a wide variety of activity through various media forms may lead to *portalisation* (Cormode and Krishnamurti 2008, 1). This implies that sites try to minimise reasons for leaving the site through detecting users' browsing habits. Based on this, a broad range of intra-site features and other media features like blogging, email, search, hosting of photos and video are provided. The Facebook API allows for a number of features to be imported to profile pages. That an increasing share of users' online communication takes place through the portal of one social network may lead to 'balkanisation', making it into a network within the Internet. Such intra-network communication is not reachable from other social networks, nor searchable with search engines.

Users do not use Facebook 'for free'. The relationship between users and the company is both of economic and social character. By spending time, attention and activity on Facebook, by exposing themselves to ads and information-processing algorithms, every user contributes to the company's profit. Since there is no money involved from users' hands, the transaction is not considered as such, only as social activity. There are some user terms to 'OK', but no labour contract. The peculiar

DOI: 10.1057/9781137446466.0007

situation here is that there is no abstract labour. Information goods are produced through social activity (attention and interaction). From this, the forces of production (computers, the Internet, clever software and big data) aggregate this in various ways and make externalities become evident for advertisers. Big data are translated into big money. The 'prosumer' (Toffler 1980), the users-as-producers, has now lifted portions of their sociality and knowledge-production into systems of production *and* consumption. There is little use in withholding the distinction between forces and relations of production in this world. Production is based on consumption and vice versa. Commodities are extracted from lifeworld practices of friendship, love, trust and social protest, and reversibly, the lifeworld is reproduced through social media.

The problem of trust

As Field et al. (2000) point out, the concept of social capital also rein-serts issues of value into social and media research. Trust, reciprocity and norms are all central concerns for adequate action. However, in spite of Putnam's and partly Coleman's emphasis on community and solidar-ity, the interest in social networks and social capital also bear witness to a crisis of the term 'community'. This term has traditionally signified a value-based collective of consensus building, solidarity and loyalty. As Fernback (2007) notes, from early sociology and social anthropology, the notion of strong, place-based solidarity through rituals has dominated.

Certainly, this notion is of relevance for certain organised groups and associations. However, for the practices of social media, the term suggests too much fellowship and cohesiveness, while saying little about the descriptive nature of interaction and communication. An extended inter-pretation is needed that (a) keeps an eye on the media dimension without discriminating ontologically between online and offline interaction, (b) that accounts for the building and transformations of multiple and overlapping networks, without losing sight of the reciprocity and stability of everyday interaction and (c) that accounts for both purposive/instru-mental, affective and expressive practices. As Castells (2000) and others have noted, the ritualistic and cultural dimension should be toned down. The point here is that one should not normatively assume cohesiveness as an a priori value of communities. While Fernbeck (2007) attempts to

DOI: 10.1057/9781137446466.0007

compensate this with ideas from Symbolic interactionism, insights from network analysis are an alternative. While a social community consists of a group of people interacting over time within a common frame of reference, social networks of the kind we study are more variable in terms of interaction intensity. By using the term network, we also intend to emphasise *the nature of relationships* more than individuals themselves.

The individuals are not so much influenced by norms and knowledge as that depend on multidirectional relationships. The common arrangements of both practical and normative kinds are less cohesive than in community-oriented groups. Compensatory mechanisms that may ensure stability of social relationships in spite of distances, mobility, heterogeneity and other risks are at play. More than previously, a lot depends on the communication context and infrastructure of the networks. Among other factors, the user interface of personal media may be seen as a functional equivalent for the normative cohesion of face-to-face settings. This indicates the pivotal role of media technologies, as normative cohesion evaporates. Only in the last decade, a wide range of personal media using both the Internet and the telephone network have been adopted by the majority of the populations in OECD countries, precisely because of increasing transparency, reach, simplicity, flexibility, adaptability and so on. The 'domestication' process analysed by Silverstone and others has proceeded towards 'personalisation' both on the web and the phone platforms. Media have lowered their threshold for use, and by this have approached the threshold for communication as such. They have become personalised by narrowing the gap between the idea of communicating with someone and actually applying a technical device to do so.

Another common term is *Communities of practice* (Wenger 1998), which refers to social networks that are engaging in similar activities and learn from each other along the way. The term and its following theoretical insights are for the most part applied in professional contexts such as working places and schools, but the term may also apply to groups of friends and families. It is a term that draws on social capital, as it refers to informal learning as a consequence of collective activities, such as argumentation and discussion, giving feedback, sharing facts and so on in communities. Families turn into communities of practice when they work in the garden or learn a new card game and when people meet to help a friend move to a new flat.

The web provides an efficient way to get in touch with not only weak ties like a plumber or the sports club, but also strong ties like old school

DOI: 10.1057/9781137446466.0007

friends. The mobile is indispensable for keeping daily contact with friends and family. Importantly, these new channels for social ties tend to complement other forms of social contact (Hampton and Wellman 1999, Wellman et al. 2003, see Agre 2001). More and more social groups tend to communicate through a broader range of media in addition to face-to-face encounters, such as email, phone, discussion-lists and so on. Networks make themselves more flexible and robust by the appropriation of the web, email, social media and so on.

Trust as a dimension of social capital is essential not only in Coleman, Putnam, Fukuyama but also in Bourdieu. As an equivalent of the lack of demand for explicit instrumental information in expressive information, an atmosphere of mutual trust must emanate from the network. Thus trust, as a platform for motivation, is more required in friendship networks than in organisation networks.

According to Fukuyama (1999), social capital is 'an instantiated informal norm that promotes cooperation between two or more individuals. The norm that constitutes social capital can range from a norm of reciprocity between two friends, all the way up to complex and elaborately articulated doctrines like Christianity and Confucianism. They must be instantiated in an actual human relationship: the norm of reciprocity exists in potential in my dealings with all people, but is actualized only in my dealings with my friends'. By this definition, trust, community and civility are fundamental. Generalising a certain amount of trust in strong and weak ties enables an undisturbed exchange of information. The argument about trust may in certain instances be overstated. In networks of weak ties, a bottom line of confidence is needed, but the real reproducing driving force is much more the actual results from being a member of the network – what is actually learnt through the status as a node. Trust is a variable that is mobilised quite differently in different sorts of networks – independently of the online–offline distinction.

Trust is a belief or an attitude towards strangers, which indicates that they are similar to oneself in terms of values, that they are friendly. This makes people less hesitant or afraid to get in touch with strangers offline or online. They would be less worried about interacting with new acquaintances. However, trust is only *one* personal factor that influences communication with strangers, and a hypothesis is that it will have only moderate influence on online behaviour. There are also factors that may compensate for lack of trust.

DOI: 10.1057/9781137446466.0007

Resource distribution

I have so far not touched upon John Urry's concept of network capital, although it resembles social capital (Urry and Elliott 2010). Urry's argument is that capital in the field of what he calls 'mobilities' is of increasing importance in social stratification, at least in the rich parts of the world. Mobility is central for the individual to reach goals and to avoid misfortune. Network capital involves a number of resources that are a prerequisite to life in the rich North (Urry and Elliott 2010). Network capital enables mobilities, which may engender and sustain social relations with distant people and which generate emotional, financial and practical benefit. More explicitly than other scholars of social capital, Urry emphasises physical mobility and transport, however, fundamentally dependent on communication technologies for coordination.

Urry's most important contribution, however, concerns the role of network capital on widening regional- and class-based divides. As is the case with other forms of capital, network capital tends to reproduce and reinforce social differences (Urry and Elliott 2010). That there is a 'Matthew law effect' regarding social capital has been confirmed by most research in the area that I have referred to above. Groups that enjoy much network capital benefit from the making and remaking of social connections. Resources tend to generate more resources. As personal media in general, and social media in particular, are used by the majority of the population in rich countries, the share of passive users with little or no motivation in generating social capital through their media use are likely to increase. Social media are dedicated to the accumulation of social capital. This is most easy to see in sites like LinkedIn, which is oriented towards career-stimulating connections, but also evident in Facebook, where the benefits are many: friendship, specialist knowledge (re)discoveries of memories, sharing of images, as well as political campaigning. The potentials are astonishing but very unevenly exploited by users. Further research is needed to identify not only the social implications of social networks, but also the necessary social conditions among users that stimulate social capital accumulation.

To conclude, unlike certain studies of domestication (see Kraut et al. 2006), social capital is here not simply viewed as social 'effects' and 'impacts' of digital media on everyday life, but more carefully concerning the implications and significances of social networks. Personal media, through processes of personalisation, become embedded in networks

DOI: 10.1057/9781137446466.0007

where effects and impacts cannot be distinguished from their causes. Rather, of interest for social research should be changes in everyday life with emphasis on its communication networks, which may have implications for individual life qualities as well as for civil life and social change (see Wittel 2001). In this, the media will play a role as structural resource. There is plenty to say about the mobile and computer as communication devices, as privacy zones, as socialisation spheres, as network facilitators, and as infrastructure for the public sphere and for social integration in general. But they cannot be empirically isolated from shifting norms, living conditions, education, family patterns, habits and a number of other variables that also undergo change. In observing communication in daily life situations, the medium is only one aspect of the message.

In social capital studies, individuals tend to have both weak and often widely scattered ties and strong ties within closed networks. At the outset, both the bonding and bridging approaches, developed by Coleman, Putnam, Granovetter and Burt, need to be represented. Some relations enable the individuals to receive trust and support on intimate matters, while weaker ties are beneficial in terms of receiving information and sharing political or moral concerns. In all cases, the individual receives collective benefits in a wide sense, which unintentionally and intentionally contribute to identity and meaning-making, which harmonises realistically with given external possibilities. And yet, the concept of social capital needs refining in order to serve as a useful tool for research on mediated social relationships in daily life. Rather than assuming cohesive civic communities and dense family bonds, the starting point for research on social capital ought to open for diverse and rather loose, ego-centric networks, in part due to geographically disperse contacts. Particularly, it is of interest to explore whether the mastering of digital media gives new possibilities to individuals who do not have much social capital through their networks, or whether they simply facilitate existing networks among those who already are 'social capitalists'. Clearly such questions are important when digital media become widely applied channels for not only friendship and trust, but also for learning, civil organisation and social change.

DOI: 10.1057/9781137446466.0007

Bibliography

Adam, F. & Ronkevik, B. (2003) 'Social capital: Recent debates and research trends'. *Social Science Information* 42(2): 155–183.

Adamic, L. A. & Adar, E. (2003) 'Friends and neighbours on the web'. *Social Networks* 25: 211–230.

Agar, J. (2003) *Constant Touch. A Global History of the Mobile Phone.* Cambridge: Icon books.

Agre, P. (2004) 'The practical republic: Social skills and the progress of citizenship', in A. Andrew Feenberg & D. D. Barney (eds) *Community in the Digital Age,* Rowman and Littlefield.

Anderson, B., Brynin, M., Gershuny, J. & Rabian, Y. (eds) (2012) *Information and Communication Technologies in Society. E-living in Digital Europe.* London: Routledge.

Aronson, S. (1971) 'The sociology of the telephone'. *International Journal of Comparative Sociology* 12: 153–167.

Bakardjieva, M. (2006) 'Domestication running wild. From the moral economy of the household to the mores of a culture', in T. Berker, M. Hartmann, Y. Punie & K. Ward (eds) *Domestication of Media and Technology,* pp. 62–79. Maidenhead: Open University Press.

Bankston, C. L. & Zhou, Z. (2002) 'Social capital as process: The meanings and problems of a theoretical metaphor'. *Sociological Inquiry* 72(2): 285–317.

Barnes, Stuart J. & Sid, Huff (2003) 'Rising sun: iMode and the wireless internet'. *Communications of the ACM* 46(11): 79–84.

Baron, S., Field, J. & Schuller, T. (2000) *Social Capital. Critical Perspectives.* Oxford: Oxford University Press.

Bateson, G. (1979) *Mind and Nature: A Necessary Unity*. New York: E. P. Dutton.

Baudrillard, J. (1983) *Simulations*. New York: Semiotext(e).

Baudrillard, J. (1994) *Simulacra and Simulation*. Ann Arbor: The University of Michigan Press.

Bauman, Z. (2000) *Liquid Modernity*. Cambridge: Polity Press.

Baym, N. K. & Burnett, R. (2008) *Amateur Experts: International Fan Labor in Swedish Independent Music*. Paper prepared for Internet Research 9.0, Copenhagen, Denmark.

Baym, N. (2010) *Personal Connections in the Digital Age*. Cambridge: Polity.

Beck, U. & Beck-Gernsheim, E. (2002) *Individualization*. London: Sage.

Beck, U., Giddens, A. & Lash, S. (1994) *Reflexive Modernization*. Cambridge: Polity Press.

Becker, B. & Wehner, J. (2001) 'Electonic networks and civil society: Reflections on structural changes in the public sphere', in Ess, Charles (ed.) *Culture, Technology, Communication. Towards an Intercultural Global Village*. Albany: State University of New York Press.

Bell, D. & Hollows, J. (2005) *Ordinary Lifestyles: Popular Media, Consumption and Taste*. Maidenhead: Open University Press.

Benkler, Y. (2006) *The Wealth of Networks. How Social Production Transforms Markets and Freedom*. New Haven: Yale University Press.

Berkenkotter, C. & Huckin, T. N. (1993) 'Rethinking genre from a sociocognitive perspective'. *Written Communication* 10(4): 475–509.

Berker, T., Hartmann, M., Punie, Y. & Ward, K. J. (eds) (2006) *Domestication of Media and Technology*. Maidenhead: Open University Press.

Bernstein, R. (ed.) (1985) *Habermas and Modernity*. Oxford: Blackwell.

Blokland, T. (2003) *Urban Bonds: Social Relationships in an Inner City Neighbourhood*. Cambridge: Polity Press.

Bolter, J. D. (1991) *Writing Space. The Computer, Hypertext, and the History of Writing*. Hillsdale: LEA.

Bolter, J. D. & Grusin, R. (1999) *Remediation. Understanding New Media*. Cambridge, MA: The MIT Press.

Bourdieu, P. (1984) *Distinction. A Social Critique of the Judgement of Taste*. London: Routledge and Kegan Paul.

Bourdieu, P. (1986) 'The forms of capital', in J. Richardson (ed.) *Handbook of Theory and Research for the Sociology of Education*. New York: Greenwood.

DOI: 10.1057/9781137446466.0008

Boyd, D. (2006) 'Friends, friendsters, and top 8: Writing community into being on social network sites'. *First Monday* 11(12).

Boy, D. (2011) 'Social network sites as network publics: Affordances, dynamics, and implications', in Z. Papacharissi (ed.) *A Networked Self. Identity, Community, and Culture on Social Network Sites*. New York: Routledge.

Brake, D. (2007) 'Personlige webloggere og deres publikum: Hvem tror de egentlig at de snakker med?', in M. Lüders, L. Prøitz & T. Rasmussen (eds) *Personlige Medier: Livet Mellom Skjermene*, pp. 141–163. Oslo: Gyldendal.

Brandtzæg, P. B. and Stav, B. H. (2004) Barn og unges skravling på nettet – sosial støtte i Cyberspace? [Social support in Cyberspace among children and young people?] *Tidsskrift for ungdomsforskning* 4(1) 27–47.

Brody, F. (2000) 'The medium is the memory', in P. Lunenfeld (ed.) *The Digital Dialectic. New Essays on New Media*. Cambridge, MA: The MIT Press.

Burt, R. (1992) *Structural Holes: The Social Structure of Competition*. Cambridge, MA: Harvard University Press.

Burt, R. S. (2000) 'The network of social capital', in R. I. Sutton & B. M. Staw (eds) *Research in Organisational Behavior*. 22 Greenwich, CT: JAI Press.

Calhoun, C. (1995) *Critical Social Theory*. Oxford: Blackwell.

Cardon, D. & Granjon F. (2004) 'Social networks and cultural practices. A case study of young avid screen users in France'. *Social Networks* 27: 301–315.

Casalegno, F. (2006) 'Connected memories in the networked digital era: A moving paradigm', in P. Purcell (ed.) *Networked Neighbourhoods: The Connected Community in Context*. London: Springer.

Casey, E. S. (2000) *Remembering: A Phenomenological Study*. Indiana University Press.

Castells, M. (1997) *The power of identity*. Oxford: Blackwell.

Castells, M. (2000) 'Toward a sociology of the network society'. *Contemporary Sociology* 29(5): 693–699.

Castells, M. (2009) *Communication Power*. Oxford: Oxford University Press.

Castells, M. (ed.) (2004) *The Network Society. A Cross-cultural Perspective*. Cheltenham: Edward Elgar.

DOI: 10.1057/9781137446466.0008

Chambers, D. (2006) *New Social Ties. Contemporary Connections in a Fragmented Society.* Houndmills: Palgrave.

Chambers, D. (2012) *A Sociology of Family Life. Change and Diversity in Intimate Relations.* Cambridge: Polity Press.

Chambers, D. (2013) *Social Media and Personal Relationships. Online Intimacies and Networked Friendship.* Houndmills, Basingstoke: Palgrave Macmillan.

Chandler, D. (1997) *Writing Oneself in Cyberspace.* Retrieved April 2014 http://www.aber.ac.uk/media/Documents/short/webident.html

Cherny, L. (1999) *Conversation and Community: Chat in a Virtual World.* New York: CSLI Publications.

Chesbrough, H., Vanhaverbeke, W. & West, J. (2005) *Open Innovation. Researching a New Paradigm.* Oxford: Oxford University Press.

Choi, J. H. (2006) 'Living in cyworld: Contextualising cy-ties in South-Korea', in A. Bruns & J. Jacobs (eds) *Use of Blogs (Digital Formations),* pp. 173–186. New York: Peter Lang.

Coleman, J. S. (1988) 'Social capital in the creation of human capital'. *The American Journal of Sociology* 94: 95–120.

Coleman, J. S. (1990) *Foundation of Social Theory.* Cambridge, MA: Harvard University Press.

Cormode, G. & Krishnamurti, B. (2008) 'Key differences between Web 1.0 and Web 2.0'. *First Monday* 13(6).

Crapeau, S. & Kretz, F. (1987) 'Methodological analysis of experiments with communication services', in L. Qvortrup et al. (eds) *Social Experiments with Information Technology and the Challenges of Innovation.* Dordrecht: D. Reidel Publ.

Crook, S. (1998) 'Minotaurs and other monsters: Everyday life in recent social theory'. *Sociology* 32(3): 523–540.

Crossley, N. (1996) *Intersubjectivity: The Fabric of Social Becoming.* London: Sage.

Cubitt, S. (2001) *Simulation in Social Theory.* London: Sage.

Dasgupta, P. & Serageldin, R. (eds) (1999) *Social Capital: A Multifaceted Perspective.* Washington, DC: World Bank.

de Certeau, M. (1984) *The Practice of Everyday Life.* Berkeley: University of California Press.

Delanty, G. (2003) *Community (key Ideas).* New York: Routledge.

Dhavan V. S., Kwak, N. & Holbert, R. L. (2001) ' "Connecting" and "disconnecting" with civic life: Patterns of internet use and the production of social capital'. *Political Communication* 18: 141–162.

DOI: 10.1057/9781137446466.0008

Döring, N. (2002) 'Personal home pages on the web: A review of research'. *Journal of Computer-Mediated Communication* 7(3).

du Gay, P. et. al. (1997) *Doing Cultural Studies: The Story of the Sony Walkman.* London: Sage.

Goffman, E. (1981) *Forms of Talk.* University of Pennsylvania Press.

Goffman, E. (1983) '"The interaction order" American Sociological Association, 1982 presidential address'. *American Sociological Review* 48(1): 1–17.

Eco, U. (1986) *Faith in Fakes: Travels in Hyperreality.* London: Minerva.

Eisenstein, E. L. (1983) *The Printing Revolution in Early Modern Europe.* Cambridge: Cambridge University Press.

Ellison, N. B., Steinfield, C. & Lampe, C. (2007) 'The benefits of Facebook "friends:" Social capital and college students' use of online social network sites'. *Journal of Computer-Mediated Communication* 12(4): 1143–1168.

Etzioni, A. (1996) 'The responsive community: A communitarioan perspective'. *American Sociological Review* 61: 1–11.

Fagerjord A. (2003) "Rhetorical Convergence. Earlier Media Influence on Web Media Form." Oslo: Faculty of the Humanities, University of Oslo.

Feenberg (1991) *Critical Theory of technology.* Oxford: Oxford University Press.

Fernback, J. (2007) 'Beyond the diluted community concept: A symbolic interactionist perspective on online social relations'. *New Media and Society* 9(1): 49–69.

Field, J., Schuller, T. & Baron, S. (2000) 'Social capital and human capital revisited', in S. Baron, J. Field & T. Schuller (eds) *Social Capital: Critical Perspectives,* pp. 243–263. Oxford: Oxford University Press.

Fisher, C. S. (1992) *America Calling: A Social History of the Telephone to 1940.* Berkeley: University of California Press.

Foucault, M. (1988) *Technologies of the Self.* Cambridge, MA: University of Massachusetts Press.

Foucault, M. (1994) 'Technologies of the self', in P. Rabinow (ed.) *Essential Works of Foucault 1954–1984.* Vol. 1. London: Penguin.

Foucault, M., (1994) *Ethics. Essential works of Foucault 1954–1984,* vol.1. P. Rabinow ed. London: Penguin

Frane, A. & Ronkevik, B. (2003) 'Social capital: Recent debates and research trends'. *Social Science Information* 42(2): 155–183.

Fukuyama, F. (1995) *Trust: The Social Virtues and the Creation of Prosperity.* New York: The Free Press.

DOI: 10.1057/9781137446466.0008

Fukuyama, F. (1999) 'Social capital and Civil Society'. IMF conference on Second generation Reforms. http://www.imf.org/external/pubs/ft/seminar/1999/reforms/fukuyama.htm#l

Garton, L., Haythornthwaite, C.& Wellman, B. (1997) 'Studying online social networks'. *Journal of Computer-Mediated Communication* 3(1).

Gershuny, J. I. (2003) *Changing Times: Work and Leisure in Postindustrial Society*. Oxford University Press.

Gershuny, J. (2003) *Time, through the Lifecourse, in the Family*. Institute for Social and Economic Research, University of Essex.

Gershuny, J. (2003). 'Web use and net nerds: A neofunctionalist analysis of the impact of information technology in the home'. *Social Forces* 82(1): 141–168.

Geser, H. (2004) *Towards a Sociological Theory of the Mobile Phone*. University of Zurich. http://socio.ch.mobile/t_geserl.htm.

Giddens, A. (1991) *Modernity and Self-Identity*. Cambridge: Polity Press.

Goffman, E. (1959) *The Presentation of Self in Everyday Life*. NY: Doubleday Anchor.

Goffman, E. (1967) *Interaction Ritual: Essays on face-to-face interaction*. Oxford: Aldine.

Granovetter, M. (1973) 'The strength of weak ties'. *American Journal of Sociology* 78: 1360–1380.

Granovetter, M. (1978) 'Threshold models of collective behavior'. *The American Journal of Sociology* 83(6): 1420–1443.

Granovetter, M. (1983) 'The strength of weak ties: A network theory revisited'. *Sociological Theory* 1: 201–233.

Granovetter, M. & Swedberg, R. (2001) *The Sociology of Economic Life*. Boulder, CO: Westview Press.

Green, N. (2002) 'On the move: Technology, mobility and the mediation of social time and space'. *The Information Society* 18: 281–292.

Gross, N. (2005) 'The detraditionalization of intimacy reconsidered'. *Sociological Theory* 23(3): 286–311.

Habermas, J. (1984) *The Theory of Communicative Action*, Vol. I. London: Heinemann.

Habermas, J. (1987) The *Philosophical Discourse of Modernity*. Cambridge, Mass.: The MIT Press.

Haddon, L. (ed.) (2005) *Everyday Innovators: Researching the Role of Users in Shaping ICTs* (Vol. 32). London: Springer.

Hadot, P. (2000) 'Reflections on the idea of "cultivation of the self"', in P. duGay, J. Evans, & P. Redman (eds) *Identity: A Reader*, pp. 373–379. London: Sage.

DOI: 10.1057/9781137446466.0008

Hampton, K. N. & Wellman, B. (1999) 'Netville online and offline: Observing and surveying a wired suburb'. *American Behavioral Scientist* 45(3): 477–496.

Harper, R., Palen, L. & Taylor, A. S. (eds) (2005) *The Inside Text: Social, Cultural and Design Perspectives on SMS* (vol. 4). London: Springer.

Harrison, T. (2009) 'Wielding new media in Web 2.0: exploring the history of engagement with the collaborative construction of media products'. *New Media & Society* 11(1–2): 155–178.

Haythornthwaite, C. (1998). "A social network study of the growth of community among distance learners." *Information Research*, 4 (1).

Haythornthwaite, C. (2000) 'Online personal networks', *New Media and Society* 2(2): 195–226.

Haythornthwaite C. (2002) "Strong, Weak and latent ties and the impact of new media." *Information Society* 18, 385–401.

Haythornthwaite C. and Wellmann B. (eds) (2002) *The Internet in Everyday Life* Malden MA: Blackwell.

Haythornthwaite, C. (2002) 'Building social networks via computer networks: Creating and sustaining distributed learning communities', in K. A. Renninger & W. Shumar (eds) *Building Virtual Communities: Learning and Change in Cyberspace*, pp. 159–190. Cambridge, UK: Cambridge University Press.

Herring, S. C., Scheidt, L. A., Wright, E. & Bonus, S. (2005) 'Weblogs as a bridging genre'. *Information Technology and People* 18(2): 142–171.

Hess, C. & Ostrom, E. (eds) (2007) *Understanding Knowledge as a Commons: From Theory to Practice*. Cambridge, MA: The MIT Press.

Hetcher, S (2009) "Hume's Penguin, or, Yochai Benkler & the Nature of Peer Production." *Vanderbilt Journal of Entertainment and Technology Law* 11(4), 995–1000.

Highmore, B. (2001) *Everyday Life and Cultural Theory*. London: Routledge.

Hodkinson, P. (2007) 'Interactive online journals and individualisation'. *New Media and Society* 9(4): 625–650.

Horst, H. A. & Miller, D. (2006) *The Cell Phone. An Anthropology of Communication*. Oxford: Berg.

Huffaker, D. A. & Calvert, S. L. (2005) 'Gender, identity, and language use in teenage blogs'. *Journal of Computer-Mediated Communication* 10(2). URL (consulted 7 March 2009): http://jcmc.indiana.edu/vol10/issue2/huffaker.html.

DOI: 10.1057/9781137446466.0008

Hui, J., Cashman, T. & Deacon, T. (2008) 'Bateson's method: Double description. What is it? How does it work? What do we learn?', in J. Hofmeyer (ed.) *A Legacy for Living Systems. Gregory Bateson as Precursor to Biosemiotics.* Copenhagen: Springer.

Huysman, M. & Wulf, V. (eds) (2004) *Social Capital and Information Technology.* Cambridge, MA: The MIT Press.

Ihde, D. (1990) *Technology and the Lifeworld. From Garden to Earth.* Bloomington: Indiana University Press.

Ihlström, C. & Henfridsson, O. (2005) 'Online newspapers in Scandinavia. A longitudinal study of genre change and interdependency'. *Information Technology and People* (2): 172–192.

Illouz, E. (2006) *Cold Intimacies: The Making of Emotional Capitalism.* Cambridge: Polity Press.

in Japanese Life. Cambridge, Mass.: The MIT Press.

Innis, H. (1951) *The Bias of Communication.* Toronto: University of Toronto Press.

Innis, H. (1986) *Empire of Communication.* Victoria: Press Porcepic.

Ito, M., Okabe, D. & Matsuda, M. (2006) *Personal, Portable, Pedestrian: Mobile Phones in Japanese Life.* Cambridge, MA: The MIT Press.

Jamieson, L. (1999) 'Intimacy transformed? A critical look at the "Pure Relationship"'. *Sociology* 33(3): 477–494.

Jarrett, K. (2008) 'Interactivity is evil! A critical investigation of Web 2.0'. *First Monday* 13(3).

Jauss, H. R. ([1967] 1982) *Toward an Aesthetic of Reception.* Minneapolis: University of Minnesota Press.

Johns, A. (1998) *The Nature of the Book.* Chicago: The University of Chicago Press.

Kasesniemi, E.-L. (2003) *Mobile Message: Young People and a New Communication Culture.* Tampere: Tampere University Press.

Kavanaugh, A. L., Reese, C., Carrol, J. M. & Rosson, M. B. (2005) 'Weak ties in network communities'. *The Information Society* 21: 119–131.

Kitzmann, A. (2004) *Saved from Oblivion: Documenting the Daily from Diaries to Web Cams.* New York: Peter Lang.

Knorr Cetina, K. (2001) 'Post social relations: Theorizing sociality in a post social environment'. *Handbook of Social Theory* 520–537.

Kohiyama (2005) 'A decade in the development of mobile communication in Japan (1993–2002)', in M. Ito, D. Okabe & M. Matsuba (eds) (2005) *Personal, Portable and Pedestrian: Mobile Phones in Japanese Life.* Cambridge, MA: MIT Press.

DOI: 10.1057/9781137446466.0008

Kraut, R., Brynin, M. & Kiesler, S. (2006) *Computers, Phones and the Internet. Domesticating Information Technology.* Oxford: Oxford University Press.

Kraut, R., Kiesler, S., Boneva, B., Cummings, J., Helgeson, V. & Crawford, A. (2002) 'Internet paradox revisited'. *Journal of Social Issues* 58(1): 49–74.

Küchler, S. & Melion, W. (eds) (1990) *Images of Memory: On remembering and Representation.*

Kuhn, A. ([1995] 2002) *Family Secrets: Acts of Memory and Imagination.* London: Verso.

Langer, S. (1942) *Philosophy in a New Key. A Study in the Symbolism of Reason, Rite, and Art.* Boston: Harvard University Press.

Lefebvre, H. (1991) *Critique of Everyday Life:* Vol. 1. London: Verso.

Levi, M. (1996) 'Social and unsocial capital: A review essay of Putnam's making democracy work'. *Politics and Society* 24(1): 45–55.

Licoppe, C. & Smoreda, Z. (2005) 'Are social networks technologically embedded? How networks are changing today with changes in communication technology'. *Social Networks* 27: 317–335.

Licoppe, C. & Smoreda, Z. (2006) 'Rhythms and ties: Toward a pragmatics of technologically mediated sociability', in R. Kraut, M. Brynin & S. Kiesler (eds) *Computers, Phones, and the Internet: Domesticating Information Technologies*, pp. 296–324. Oxford: Oxford University Press.

Liestøl, G. (2009) 'Notes on mobility, localization & the possibility of genre design', in I. Wagner et al. (eds) *Exploring Digital Design.* New York: Kluwer Academic/Plenum.

Liestøl, G. (2011) 'Situated simulations between virtual reality and mobile augmented reality: Designing a narrative space', in B. Furht (ed.) *Handbook of Augmented Reality.* Springer.

Liestøl, G., Rasmussen, T. & Stenarson, T. (2011) 'Mobile innovation: Designing & evaluatiing situated simulations', in *Digital Creativity* 22(3), pp. 172–184. Abingdon: Routledge, Taylor & Francis Group (2011).

Lin, N. (2001) *Social Capital. A Theory of Social Structure and Action.* Cambridge: Cambridge University Press.

Ling, R. (2004) *The Mobile Connection: The Cell Phone's Impact on Society.* San Fransisco: Elsevier.

Ling, R. (2008) *New Tech, New Ties. How Mobile Communication is Reshaping Social Cohesion.* Cambridge, MA: The MIT Press.

Ling, R. & Haddon, L. (2001) 'Mobile telephony and the coordination of mobility in everyday life'. *Telenor FoU*, R16.

DOI: 10.1057/9781137446466.0008

Livingstone, S. (2002) *Young People and the New Media. Childhood and the Changing Media Environment*. London: Sage.

Lüders, M. (2007) *Being in Mediated Spaces. An Enquiry into Personal Media Practices*. PhD-thesis. Oslo: University of Oslo.

Lüders, M. (2008) 'Conceptualising personal media'. *New Media and Society* 10(5): 683–702.

Lüders, M., Prøitz, L. & Rasmussen, T. (2010) 'Emerging personal media genres', in *New Media and Society* 12(6): 1–17.

Luhmann, N. (1990) *Essays on Self-reference*. Stanford: Stanford University Press.

Luhmann, N. (1996/2000) *The Reality of the Mass Media*. Stanford: Stanford University Press.

Luhmann, N. (1998) *Observations on Modernity*. Stanford: Stanford University Press.

Luhmann, N. (2012) *Introduction to Systems Theory*. NJ: Wiley.

Madianou, M. & Miller, D. (2013) 'Polymedia: Towards a new theory of digital media in interpersonal communication'. *International Journal of Cultural Studies* 16(2): 169–187.

Manovich, L. (2001) *The Language of New Media*. Cambridge, MA: The MIT Press.

Martin, J. R. (1989) *Factual Writing: Exploring and Challenging Social Reality*. Oxford: Oxford University Press.

Martin, J. R. (1992) *English Text: System and Structure*. Philadelphia/ Amsterdam.

Matsuda, M. (2005) 'Discourses of *Keitai* in Japan', in M. Ito, D. Okabe & M. Matsuba (eds) *Personal, Portable and Pedestrian: Mobile Phones in Japanese Life*. Cambridge, MA: The MIT Press.

McLuhan, M. (1964) *Understanding Media: The Extensions of Man*. New York: Signet.

McLuhan, M. & McLuhan, E. (1988) *Laws of Media: The New Science*. Toronto: University of Toronto Press.

Meyrowitz, J. (1986) *No Sense of Place*. Cambridge, MA: The MIT Press.

Misztal, B. A. (1996) *Trust in Modern Societies. A Search for the Base of Social Order*. Cambridge: Polity Press.

Mitchell, W. (2003) *Me++*. Cambridge, MA: MIT Press.

Petersen, S. M. (2008). 'Loser generated content: From participation to exploitation'. *First Monday* 13(3).

DOI: 10.1057/9781137446466.0008

Naphapiet, J. & Ghosal, S. (1998) 'Social capital, intellectual capital and organisational advantage'. Management research papers 97/6. Oxford: Oxford Centre for Management Studies.

Norris, P. (1996) 'Does television Erode social capital? A reply to Putnam'. *Political Science and Politics* 29: 474–479.

Okada T. (2006) 'Youth culture and the shaping of Japanese mobile media: Personalization and the *Keitai* internet as multimedia', in M. Ito, D. Okabe & M. Matsuba (eds) (2005) *Personal, Portable and Pedestrian.*

Ong, W. (1982/1991) *Orality and Literacy: The Technologizing of the Word.* London: Routledge.

O'Reilly, T. (2005) 'What is web 2.0. Design patterns and business models for the next generation of software" http://OReilly.com

Papacharissi, Z. (ed.) (2011) *A Networked Self. Identity, Community, and Culture on Social Network Sites.* New York: Routledge.

Pichault, F. (1978) 'Social experiments with IT from the initiator's point of view', in L. Qvortrup et al. (eds) *Social Experiments with Information Technology and the Challenges of Innovation.* Dordrecht: D. Reidel Publ.

Portes, A. (1998) 'Social capital: Its origins and applications in modern sociology'. *Annual Review of Sociology* 24: 1–24.

Poster, M. (1995) *The Mode of Information.* Cambridge: Polity Press.

Prøitz, L. (2005) "Cute Boys or Game Boys? The Embodyment of feminity and masculinity in young Norwegians' Text-message love projects" *Fibreculture Journal* No. 6.

Prøitz, L. (2007) 'The mobile phone turn – a study of gender, sexuality and subjectivity in young people's mobile phone practices'. University of Oslo: *Acta Humaniora*, Unipub Press.

Putnam, R. (1995) 'Tuning in, tuning out: The strange disappearance of social capital in America'. *Political Science and Politics* 28(4): 664–683.

Putnam, R. et al. (1993) *Making Democracy Work: Civic Traditions in Modern Italy.* Princeton, NJ: Princeton University Press.

Quan-Haase, A. & Wellman, B. (2004) 'How does the internet affect social capital?' in M. Huysman & V. Wulf (eds) *Social Capital and Information Technology.* Cambridge, MA: The MIT Press.

Qvortrup, L. et al. (eds) (1987) *Social Experiments with Information Technology and the Challenges of Innovation.* Dordrecht: D. Reidel Publ.

Rasmussen, T. (1996) *Communication Technologies and the Mediation of Social Life.* Oslo: The University of Oslo.

DOI: 10.1057/9781137446466.0008

Rasmussen, T. (2000) *Social Theory and Communication Technology*. Aldershot: Ashgate.

Rasmussen, T. (2003) *Luhmann. Kommunikasjon, medier, samfunn*. Bergen: Fagbokforlaget.

Rasmussen, T. (2007) *Nettverksformelen. Hvordan det sosiale livet henger sammen*. Oslo: Unpub.

Rasmussen, T. (2003) 'On distributed society: The internet as a guide to a sociological understanding of communication', in G. Liestøl, A. Moorison & T. Rasmussen (eds) *Digital Media Revisited. Theoretical and Conceptual Innovations in Digital Domains*. Cambridge, MA: The MIT Press.

Rheingold, H. (2003) *Smart Mobs: The Next Social Revolution*. Berkeley: Perseus Publishing.

Sandbothe, M. (2000) 'Interactivity, hypertextuality – transversality. A media-philosophical analysis of the Internet'. *Hermes, Journal of Linguistics* (24): 81–102.

Schuller, T., Baron, S. & Field, J. (2000) 'Social capital: A review and a critique', in S. Baron, J. Field & T. Schuller (eds) *Social Capital. Critical Perspectives*. Oxford: Oxford University Press.

Sennett, R. (2004) *Respect. The Formation of Character in an Age of Inequality*. London: Penguin Books.

Silverstone, R. & Haddon, L. (1996) 'Design and domestication of information and communication technologies: Technical change and everyday life', in Mansell & Silverstone (eds) *Communication by Design. The Politics of Information and Communication Technologies*. Oxford: Oxford University Press.

Ithiel de Sola, P. O. O. L. (1983) *Technologies of Freedom*. Harvard University Press.

Spigel, L. (1992) *Make Room for TV: Television and the Family Ideal in Postwar America*. Chicago: University of Chicago Press.

Sproull, L. & Kiesler, S. (1991) *Connections: New Ways of Working in the Networked Organization*. Cambridge, MA: MIT Press.

Tanis, M. & Postmes, T. (2003) 'Social cues and impression formation in CMC'. *Journal of Communication* 53(4): 676–693.

Toomi, I. (2002) *Networks of Innovation: Change and Meaning in the Age of the Internet*. Oxford: Oxford University Press.

Tribble, E. B. (1993) *Margins and Marginality: The Printed Page in Early Modern England*. University of Virginia Press.

DOI: 10.1057/9781137446466.0008

Turkle, S. (1997) *Life on the Screen: Identity in the Age of Internet.* New York: Touchstone.

Turkle, S. (1999) ' "Cybespace and identity" symposium'. *Contemporary Sociology Journal of Reviews* 28(6).

Turkle, S. (2011) *Alone Together. Why We Expect More from Technology and Less from Each Other.* New York: Basic Books.

Urry, J. & Elliott, J. (2010) *Mobile Lives.* London: Routledge.

Uslaner, E. M. (2004) 'Trust, civic engagement, and the internet'. *Political Communication* 21(2): 223–242.

van den Hooff, B. & De Ridder, J. A. (2004) 'Knowledge sharing in context: The influence of organizational commitment, communication climate and CMC use on knowledge sharing'. *Journal of Knowledge Management* 8(6): 117–130.

Watzlawick, P. et al. (1967) *Pragmatics of Human Communication: A Study of Interactional Patterns, Pathologies and Paradoxes.* New York: W.W. Norton.

Weber, M. (1958) 'From Max Weber. essays in sociology', in H. H. Gerth & C. W. Mills (eds). New York: Oxford University Press.

Webster, F. (ed.) (2001) *Culture and Politics in the Information Age. A New Politics?* London: Routledge.

Wellman, B. (1999) *Networks in the Global Village.* Boulder, CO: Westview Press.

Wellman, B. & Hampton, K. (1999) 'Living networked on and offline'. *Contemporary Sociology* 28(6): 648–654.

Wellman, B. & Haythorntwaite, C. (eds) (2002) *The Internet in Everyday Life.* Oxford: Blackwell.

Wellman, B. et al. (2001) 'Does the Internet increase, decrease or supplement social capital?: Social networks, participation and community commitment'. *American Behavioral Scientist* 45(3): 437–456.

Wenger, E. (1998) *Communities of Practice: Learning, Meaning, and Identity.* Cambridge: Cambridge University Press.

White, H. (1965/2008) 'White notes on the constituents of social structure'. *Sociologica* No. 1.

Willensky, J. (2006) *The Access Principle. The Case for Open Access to Research and Scholarship.* Cambridge, MA: The MIT Press.

Wittel, A. (2001) 'Toward a network sociality', *Theory, Culture & Society* 18(6): 51–76.

Wittgenstein, L. (1953) *Philosophical Investigations.* Oxford: Blackwell.

DOI: 10.1057/9781137446466.0008

Index

DOI: 10.1057/9781137446466.0009